Game Audio with FMOD and Unity

T0133654

Game Audio with FMOD and Unity introduces readers to the principles and practice of game audio through the process of creating their own First Person Shooter (FPS) game.

All the basics are covered, as well as a simple introduction to coding. Using the free software Unity and FMOD Audio Middleware, the reader will be able to create a game of their own and develop a portfolio that demonstrates their capacities in interactive sound design.

Perfect for classroom use or independent study, *Game Audio with FMOD and Unity* also comes with a full suite of audio assets provided on a companion website.

Ciarán Robinson has over 20 years' experience in teaching Audio Production, and has developed and delivered degree courses for universities and colleges. His original background is in TV, film, music, and studio electronics, as well as in video production and editing. For the past three years he has worked in online education, where he developed courses on Game Audio, as well as working with industry professionals to produce courses on Film Making and Visual Effects. Ciarán has written technical articles and equipment reviews for several magazines and websites, and currently runs a small video production studio, where he creates training and tutorial videos.

Game Audio with FMOD and Unity

Ciarán Robinson

Routledge
Taylor & Francis Group

NEW YORK AND LONDON

First published 2019
by Routledge
52 Vanderbilt Avenue, New York, NY 10017

and by Routledge
2 Park Square, Milton Park, Abingdon, Oxon, OX14 4RN

Routledge is an imprint of the Taylor & Francis Group, an informa business

Library of Congress Cataloging-in-Publication Data
Names: Robinson, Ciarán, author.
Title: Game audio with FMOD and Unity / Ciarán Robinson.
Description: New York, NY : Routledge, 2019. | Includes index.
Identifiers: LCCN 2018053387 (print) | LCCN 2018053712 (ebook) |
ISBN 9780429850349 (pdf) | ISBN 9780429850332 (epub) |
ISBN 9780429850325 (mobi) | ISBN 9781138315969 (hbk : alk. paper) |
ISBN 9781138315976 (pbk : alk. paper) | ISBN 9780429455971 (ebk)
Subjects: LCSH: Computer games–Programming. |
Computer sound processing. | Video games–Sound effects. |
Unity (Electronic resource)
Classification: LCC QA76.76.C672 (ebook) |
LCC QA76.76.C672 R6336 2019 (print) |
DDC 794.8/1525–dc23
LC record available at https://lccn.loc.gov/2018053387

ISBN: 978-1-138-31596-9 (hbk)
ISBN: 978-1-138-31597-6 (pbk)
ISBN: 978-0-429-45597-1 (ebk)

Typeset in Berling and Futura
by Newgen Publishing UK

Visit the companion website: www.routledge.com/cw/robinson

Contents

v

Figures

Preface

Welcome to *Game Audio with FMOD and Unity*!

Several years ago, I was asked to design a university course on game audio. I agreed – without really realizing what I was getting into! Luckily, this coincided with the change in the way Game Engines were marketed. Suddenly, all of the tools used to make games were available for free. All I had to do was to learn how to use them…

This is the book that I needed then. It introduces the concepts behind game audio, and gives you the chance to put them into practice in your own game.

It is aimed at readers with a background in audio production who wish to expand into the area of computer game sound (though it will also be useful for game makers who wish to learn more about audio design and implementation).

The video game industry is huge. It now makes nearly double the money of the film and music industries combined! As well as the obvious improvements in graphics and gameplay, audio has come a long way since the bloops and bleeps of the 1980s. Gamers now expect fully immersive sound, so there's a real need for designers who understand the requirements of games, and who have the skills to put them into practice.

Though there's some overlap with film and TV sound, game sound designers need a unique set of abilities. Sound effects and music need to be interactive, responding to changes and events in the game.

This book focuses on the design and creation of game sound, but also looks at how it is implemented. This means that we have to use code.

Coding isn't necessarily the job of the sound designer – an audio programmer has the responsibility of getting the sounds to work with the game. However, if you want to see your work in action, you're going to have to get your hands dirty!

This book provides you with all of the code you'll need to make your game, and I explain exactly what it's doing step-by-step in the "Walkthroughs." If you've no prior experience with coding, then these will be a great introduction.

If you want to focus on sound design and game making, you can skip past these sections – in fact, it's worth considering working through this book twice. The first time, concentrate on making your game, and the second time you can look more carefully at how the code works.

As well as the code resources, I've also created sound effects, music, and 3D models for you to use. You can find these at gameaudiobook. com, and it will be worth downloading all of these right now.

You don't have to just use my samples and models though! You'll get a lot more from this book if you create and record your own assets.

Once you've finished, you'll have a playable game to start off your work portfolio.

You're probably eager to get started so let's jump right in and take our first look at the software we'll using.

Game Engines

Learning Outcomes

By the end of this chapter, you will be able to:

- Define the function of a Game Engine.
- Download and install the Unity Game Engine.
- Define the function of an Integrated Development Engine (IDE).

Game Engines

It is possible to make a game from scratch. Developers can create their own tools to generate and display the graphics, and write their own code to give physical properties to objects in the game world. This has the advantage that the software is designed and optimized specifically to do what is needed. However, it takes a lot of time, money, and experience.

An alternative solution is to use a pre-made Engine. This provides you with all of the basic tools you need, so you can then spend your efforts making the actual game.

Third-party Game Engines really took off in the 1990s with Id Software's Doom and Quake. After the initial success of these games, the software behind them was licensed out to other developers as the Doom and Quake Engines.

A Game Engine would include:

- Audio Engine (provides audio control and processing for the game).
- Game Editor (allows you to arrange and configure the game world).
- Graphics Engine (renders lighting, shading, and illumination within the game world).
- Physics Engine (simulates physics in the game world).

Obviously, there are financial considerations (for example, you will have to pay the Engine developers a license fee), and the developers will have to optimize the performance specifically for the game they create.

These days there are many Engines available (many of which also started off life as the software behind other games).

Examples include:

2

- Cry Engine: Developed by Crytek (originally created for the first Far Cry game). Available on Windows only.
- Amazon Lumberyard: a version of Cry Engine, developed (unsurprisingly) by Amazon. Available on Windows only.
- GameMaker Studio: originally designed for 2D games. Uses a drag and drop interface, which is very easy to learn. Windows only.
- Frostbite: developed by (and exclusive to) EA DICE (originally for the Battlefield games).
- Unity: developed by Unity Technologies. Originally OSX only, but now also available for Windows.
- Unreal Engine: developed by Epic Games for the Unreal game. Available for Mac OSX and Windows.

The best known of these are Unity and Unreal.

You can download and use either of these for free – you can even commercially release games without having to pay the Engine developers anything (though once your project reaches a certain point, license fees and revenue splits kick in – see the specific software licenses for details). For learning purposes, this is ideal. None of the software used in this book will cost you a penny.

So what's in it for the Engine developers?

Well, if Game Makers learn using their software, they're very likely to stick with it when they graduate to making higher-budget games. In addition, both Unity and Unreal Engine have integrated shopfronts, where you can buy (and sell) assets, including graphics, plugins, and even game templates. The Engine developers take a percentage of every sale.

Both of these Game Engines are fantastic, but I've decided to base this book around Unity.

- Unity uses C# for coding, which is a relatively simple language to learn (Unreal uses "Blueprints" and C++).
- It can run on a lower-spec computer (which I've found to be very important when teaching in computer labs).

Unity and Coding Languages

We use C# for coding in Unity. However, it actually runs behind the scenes using C++. The C# code is "wrapped" by the "Unity Engine," before Unity can run it. Why? Well, C# is relatively easy to code, but C++ has better performance. This way you have the best of both worlds.

How does this affect you? Well, it doesn't – everything is taken care of behind the scenes. I just mention it for interest's sake.

Unity used to give an option for coding in Javascript (well, a special version called UnityScript), as well as another language called "Boo," but these are no longer supported (support for UnityScript was removed in Unity 2018.2).

Oddly enough, Javascript is having a slight resurgence with the development of WebGL. This is a Javascript API (Application Interface) that makes it possible to run games within a web browser, without the need for any additional plugins (it's the system behind Facebook's Gameroom). It works by actually converting the Unity C++ code into Javascript. However, not all of Unity's features are available in WebGL, and it's not (yet) very straightforward to get this working with FMOD.

3

Unity has been around since 2005. It's been used to create games such as *Cuphead*, *Superhot*, and *Super Mario Run*. It was the first big-name Game Engine to provide a free fully-functional license, and as a result, quickly became very popular in the indie game industry.

The game editing tools and presets in Unity are amazingly powerful and easy to use. It's possible to quickly put together a playable game scene, without writing or editing a single line of computer code (you'll see how to do this in Chapters 7 to 9).

However, once you start to get to grips with Unity (and after you've completed this book), I'd recommend that you spend some time playing with Unreal Engine. Much of what we'll cover can be transferred from one to the other, and if you plan to work in the industry, you should be familiar with all the different software options available.

Unity runs on both Mac OSX and Windows. I'll be using a Mac, but where there are any significant differences (for example, some of the menus are found in different places), I'll show both versions.

4

Hardware Requirements

You'll need a reasonably up-to-date computer.

- If you're using a PC, you'll need Windows 7 SP1+, 8 or 10 (64-bit versions only).
- On a Mac, you'll need OSX 10.9 or later.

Your computer will also need SSE2 instruction set support. If you're running OSX or Windows 8 (or later), you definitely have this (to be honest, any computer since 2003 probably has this).

Finally, your graphics card needs to have DX10 (shader model 4.0) capabilities. Again, almost any reasonably up-to-date card will be able to do this.

That's pretty much it. Obviously, for a complex game you'll need to max your RAM, graphics capabilities, and CPU speed, but our game is going to be quite simple. Later, we'll look at how to set the default graphics quality, so you can set it to match the best performance that your system can handle.

Installing the Software

To start with, head to unity3d.com, and look for the log in/create Unity ID menu. It's probably at the top right (at the moment it uses an outline of a head icon, but this might change by the time you're reading this).

Create a Unity ID. You'll need a valid e-mail to confirm your account – and note down your password – you'll be needing it very shortly! Now click where it says "Get Unity."

There are three versions of Unity license available – Personal, Plus, and Pro. You want Unity Personal. Unity Personal is missing a few features from the other versions (none of which are required for this book). For example:

- Unity Personal does not have the "Pro Editor Skin UI." This changes the look of Unity, giving it a darker interface (which can reduce eye strain over long use).
- Unity Personal cannot create multiplayer games with more than 20 simultaneous users.
- Unity Personal does not have Performance Reporting (this is probably the most significant difference, as it is extremely useful when optimizing the game).

Click on "Try Personal," read through and agree to the Terms of Service.

Now click on the Download Unity Installer for your platform (Mac OSX or Windows). This actually downloads the Download Assistant.

Once this finishes downloading (it's quite small, so won't take long), double-click to open and run it.

You'll see the usual terms of service menus, and once you agree to these, you'll get to the Components menu, which allows you to select exactly what you're about to install.

(If you realize later that you've missed anything out, you always can run the downloader again.)

Obviously, you want Unity, but you'll also need an IDE. You won't actually be coding directly inside Unity. Instead, you'll need an "Integrated Development Environment" – an application designed for creating and editing code.

Unity comes bundled with the Visual Studio IDE so make sure that this option is checked (until recently, the Mac version of Unity came

5

bundled with an IDE called Monodevelop, but this has been changed as Visual Studio already supports some of the new C# features planned for inclusion in Unity).

IDEs

An IDE is an Integrated Development Environment. It is an application designed specifically for writing and editing code.

You don't actually need an IDE – any application that can edit text would do the job. However, an IDE makes things much easier – for example, it can auto-fill, as well as indicate any mistakes in the "grammar" of the code.

You're not limited to using Visual Studio with Unity – alternatives include Monodevelop, Visual Studio Code, and JetBrains Rider.

To change your IDE, go to Unity > Preferences, and in the External Tools menu, change the External Script Editor setting.

6

We'll definitely be needing the Standard Assets. We'll be using these as the foundation for our game.

If you're totally new to Unity, you might want to add the Example Project, so you can do some exploring of the application of your own – but this will significantly add to the download size.

There are also options to add Build Support so you can create games for other platforms. If you're on a Mac, I'd recommend adding Windows Build Support (Windows Mono Scripting Backend) and vice versa.

Once you've decided what you need, click on "Continue," and complete the install. It'll be a few gigabytes, so you've got time to put the kettle on… Don't trash the Installer though – you might need it again at a later date (for example, if you need to add Build Support for additional platforms).

If you left the option checked, Unity will launch as soon as it's installed. Log in with the account and password you just created.

You'll now have to complete a short questionnaire, but don't worry – you only need to do this the first time you launch the software.

Finally, you'll get to the Launcher window, where you can open and create a new Unity project.

There are two Tabs available – Projects (which will be empty until you create something), and "Learn," which provides a number of free tutorials.

That's got Unity up and running. If you've got the time, it's worthwhile having a look around here, but if you want to jump straight into Chapter 2, quit Unity for now, and we'll take a look at Audio Middleware.

7

2 Audio Middleware

Learning Outcomes

By the end of this chapter, you will be able to:

- Define the function of Audio Middleware.
- Download and install FMOD.

It would be possible to implement all of the sound and music inside Unity – it has a fully featured built-in Audio Engine, which has everything you need to mix and process audio. However, to make your audio truly interactive would require an extensive background in coding.

Many sound designers have no experience with coding, so in these circumstances they would work closely with the audio programmer, who would then implement their sound and music in the game.

Another solution is to use Audio Middleware. These are applications (or plugins) that are designed specifically to create interactive sounds, which can then be integrated with the Game Engine.

There are a number of Audio Middleware options, and if you're intending a career in game audio, you should be as familiar with as many of these as possible.

Wwise (by Audiokinetic)

Wwise is probably the best known Audio Middleware. It's an incredibly powerful system, and has been used in games including Overwatch and DOOM.

However, Wwise has quite a steep learning curve, and (what seems at first) a very intimidating and confusing interface – it resembles a spreadsheet more than a Digital Audio Workstation (DAW).

If you wish to become a professional game sound designer, I'd definitely recommend learning Wwise, as it's extremely common in the industry. If you were to jump straight into learning Wwise, you'd likely find it very difficult. However, once you've worked through this book, you'll be familiar with the concepts behind game audio, and it will start to make a lot more sense.

Audiokinetic provide a free version of Wwise that allows you to explore and implement all of its features (you are limited to 200 audio assets, but if you need more, you can apply for their free non-commercial license). There's also a free Wwise 101 course available.

Wwise can integrate with Unity and Unreal Engine (and third party integration is possible for CryEngine and Lumberyard). It also has integration with Steinberg's Nuendo DAW, which allows you to quickly transfer audio, MIDI and markers between the two applications. This effectively brings all of the capabilities of Nuendo into Wwise.

Fabric (by Tazman-Audio)

Fabric is not a dedicated application – it's a Unity plugin, which provides dedicated tools for building and configuring interactive audio. I wouldn't recommend Fabric to an absolute beginner, but once you're familiar with Unity and the concepts behind game audio, it will be worth spending some time experimenting with this – there's a free "Small License" available.

Elias (Adaptive Music)

Elias (apparently, "Elastic Lightweight Integrated Audio System") calls itself a "Music Engine."

9

It is expected to be used in combination with another Audio Engine that provides the sound effects (there's an inherent latency in the Elias AI, which means that it's not suitable for providing SFX).

It is designed for composers to arrange and test adaptive music. If you've used Ableton Live before, it will look somewhat familiar – music sections are arranged much like clips in the Live Session view.

Elias provide plugins for integration with Unity and Unreal Engine, as well as "bindings" that allow it to be used alongside FMOD, Wwise and Fabric. There's also a Software Development Toolkit (SDK) for integration with other Game Engines.

If you're interested in composing for games, then it's definitely worth playing around with Elias. There's a free license available (but keep in mind that this version has a limited feature set).

FMOD Studio (by Firelight Technologies)

I have chosen to base this book around FMOD Studio, as it's ideal for introducing the basic concepts and implementation of game sound. It has been designed to look familiar to anyone who's used a DAW before (though, as we'll see, it has some very different capabilities).

But don't be put off by the simple interface – FMOD is extremely powerful and versatile, and has been used in games including *FORZA Motorsport 7*, *Just Cause 3*, and *Rise of the Tomb Raider*.

Firelight provide tools for FMOD integration with Unity and Unreal Engine, and there is support for third-party audio plugins (though these will need to be licensed for use within a game). The FMOD Studio application is also free to download and use – you only need to pay a license fee once you start releasing commercial games.

Audio Middleware vs the Audio Engine

An Audio Engine provides audio control and processing for the game.

Unity has a built-in Audio Engine, but when a game is integrated with Audio Middleware, the function of this is replaced by the third-party Audio Engine.

Fabric is an exception to this – it still uses the built-in Unity Audio Engine, while providing the tools found in other Middleware solutions. This has the added advantage that games will be immediately compatible with all platforms that Unity supports.

Installing FMOD

Head to FMOD.com, and look for the Profile option (it's probably at the top right of the page). Here you'll see an option to register for an account.

Once you get through the registration process, go to the Download page. In the FMOD Studio Tab, choose which version you need (Mac OSX or Windows 64 bit), and start the download.

You'll also see an option to download the FMOD Studio API. Don't bother with this – this is the Application Processing Interface, which is only needed by the Audio Programmer.

Then head back to the FMOD website, and change to the Integrations Tab. You'll need the Unity Integration that matches the version of Unity you installed in Chapter 1.

You should now have two downloaded items – the FMOD Studio Installer, and a "UnityPackage."

Put the UnityPackage somewhere safe. This is used to integrate your Unity and FMOD projects together (we'll get around to this in Chapter 10).

Now run the installer (on a Mac, this is simply a drag-and-drop install).

And that's all it takes! You now have all nearly all the software you need installed (you'll also need a DAW). In Chapter 3, we'll be taking a look at the differences between game audio and film sound.

11

Game Audio vs Film Sound

3

Learning Outcomes

By the end of this chapter, you will be able to:

• Recognize the major differences between audio for games and audio for film.

If you have any background in film and TV sound, you will find that there is a definite skill set overlap with game audio (and if you're working on cut scenes, essentially they are the same job).

However, game audio has its own set of challenges that you should be aware of before we start:

Event Triggering

A computer game is non-linear. You can't predict when, or even if, the player will perform an action. These are not problems we have to deal with in film – every time you watch a scene, the same events will happen at the exact same time. In a game, every play-through will be different – we have no precognition.

So how do we approach this? We need to set up interactive "Events" that are triggered by the game. These are configured to respond differently, according to all the possibilities available to the player.

An Event could do something as simple as causing a sound effect to play, or it could trigger a complex sequence of actions (such as changing the mix settings, effect parameters, and music score).

In our game, we will be configuring Unity to trigger FMOD Events on actions such as:

On Collision.Enter

Triggered when one Game Object collides with another. For example, when the player collides with a Crate, a sound effect will be played.

On Trigger.Enter/Exit

Triggered when one Game Object enters/exits an area. For example, we will use this to change the mix settings when the player enters and leaves the Fort.

On Input.GetButtonDown

Triggered when a specified button is pressed. For example, when the player presses the left mouse button, a Rocket will be fired, and FMOD will play a sound effect.

On Object.Start

Triggered as soon as the Game Object is added to the scene. For example, when a rocket is instantiated (this just means that an instance of the Rocket is created), the rocket trail sound effect will be triggered.

We'll look at several other ways to trigger Events as and when they are needed in our game.

Event Variables

We also need to change the sound according to how the action happens – will the player brush against the edge of the door frame when exiting a room, or will they walk headfirst into the wall? Our sound could change to reflect this. We can use several different techniques here:

Event Conditions

We can trigger different FMOD Events according to states and conditions in the game. For example, the player's footsteps will trigger a different Event according to the surface they are walking on.

Parameter Control

We can use game variables to control the Events. For example, the velocity of an object could affect playback – perhaps the faster the object is moving, the higher the pitch?

Our First Person Shooter game doesn't have anywhere we could easily implement this, but there's a bonus chapter (Parameters 2) where we'll take a look at how this can be done.

14

Repetition

One of the core tenets of game sound is "repetition destroys immersion." If you hear the same sound again and again, then it distracts the player, and detracts from the game.

This can be particularly noticeable with dialogue (or indeed, any vocal sound effects). Games based on film licenses are particularly guilty of this. They often have limited vocal sample assets, and when you hear the same wisecrack over and over again, the repetition quickly becomes very obvious.

Although you may be tempted to throw a Wilhelm scream into your game, remember that what might seem quite funny on the first play through, will get less funny every time (yes, I'm looking at you *Star Wars Battlefront 2*…).

The Wilhelm Scream

The Wilhelm scream has been heard in hundreds of films, games, and TV programs. Originally it was part of the Warner Bros sound library (there are actually six versions of the scream), but it gets its name from the film *The Charge at Feather River*, where the Wilhelm character gets shot in the leg with an arrow.

It became a running joke to fit the scream into films, which inevitably made its way over to games. However, though you might get away with using it in a cut scene, every time the player hears it, it can break their immersion, and take them out of the game – doing the exact opposite of what you're trying to achieve!

Randomization

One approach to this is to add elements of randomization. For example, multiple samples can be placed in one Event, and when the Event is triggered, one of these samples is randomly selected and played.

One of the first examples of this we'll come across will be for our footsteps Event. We'll need several footstep samples – but not too many!

Take a quick walk across the room, and listen to the sound of your feet – there's unlikely to be much difference from one step to the next. If your footstep samples change too much, it will sound as if the game avatar is swapping shoes at every stride.

There's an art to getting this right – hitting the goldilocks zone for the right amount of variation.

We also need to consider that file space is at a premium in a game. For example, as I write this, the over-the-air size limit for iOS apps is 150MB. Any larger, and your iOS device will have to connect to a wi-fi network.

Developers try to keep their apps under this limit (if you have to wait until you connect to a wi-fi network before you can download this great game a friend tells you all about, then you're far less likely to get around to installing it).

150MB sounds like quite a lot of file space – enough for 15 minutes of uncompressed "CD quality" audio. However, this needs to be shared with all of the game assets – and graphics tend to have priority over sound… We need to get as much use as possible from every audio asset we put into the game.

For example, we could add randomization to the Event playback – we could add slight pitch and amplitude variation – not enough to be immediately noticeable, but enough to give a subtle difference every time the sample is played (we'll experiment with this with our explosion sound effects).

As an exercise, try taking some footage of a fighting game, and replacing all of the punch sound effects using only four audio files. You can layer these, change the playback speed/pitch and amplitude, but don't use any effects processing or EQ. This will give you great practice in squeezing every last drop from your audio assets.

Duration

How long is a video game? Well, a typical "Triple-A" title might come in at around 30 to 40 hours. And then you have MMORPGs (Massively Multiplayer Online Role-Playing Games) such as *World of Warcraft*, where players can spend thousands of hours in the game world.

Obviously, there isn't 40 hours of soundtrack scored and recorded for every game… One solution to this would be to loop the music. However, as anyone who has worked in retail over the holiday period will tell you, hearing the same tunes again and again quickly becomes irritating!

We can use randomization techniques to add variation to the musical arrangement, but we also need to use Events to control the music progression. This way we can use the music to inform the player, and add to the emotion of the game. For example, in *Mario Kart 8*, there are points along the course that trigger different music sections – the score always becomes more upbeat and intense as you start the last lap.

We'll keep things quite simple in our game. There will be several "pickups" scattered around the level. Each one the player finds adds to a score parameter, which controls the music arrangement.

The Music of *GTA V*

In *GTA V*, once you get in a car, you can tune your radio from a selection of stations, offering different genres of music. The DJ announcements also help build the game world, reporting on events happening in the city of Los Santos.

This almost diegetic approach works well, as listening to music while driving a car is something that goes together naturally (by the time I finished playing the game, I'd heard most of the music and DJ announcements more than once, but this issue has been fixed by offering additional DLC radio stations).

For the main missions in the game, a dedicated score that interacts with the in-game events is used. This gives the missions a more cinematic feel, and helps them stand out from the rest of the gameplay.

The interactive score was also arranged linearly, mixed, and released as "The Music of Grand Theft Auto V – Volume 2." Definitely worth a listen...

Diegetic Sound

The term "diegetic" is used to refer to sounds that originate within the game world – as opposed to "mimetic" (mimetic sounds originate outside the game world – though usually we refer to these as "non-diegetic").

Examples of diegetic sounds in our game will include:

- Explosion sound effects
- Footsteps sound effects
- Background ambience
- Sentry Drone alert and scan sound effects

In most games, narration, music and interface sound effects will be non-diegetic. Obviously there are exceptions to this – the first example that comes to mind is the music in the Fallout games. As you wander through the post-apocalyptic landscape, you often come across radios playing music by Bing Crosby and Nat King Cole.

Virtual reality offers a new set of challenges in game music. The aim of VR is immersion – the player should feel they are truly present in the game world. Anything that detracts from this should be avoided.

Now consider what it's like when you walk around listening to music on headphones... essentially, you are listening to non-diegetic music. This has the effect of distancing you from your environment.

17

But we still want to use music in the game – it is an almost essential tool for adding emotion. We're still in the early pioneering days of VR, and there isn't yet a "standard" way to do this, but one technique is to use elements of diegetic music. For example, in the Playstation VR demo street luge game, music comes from the stereos of the cars that the players pass as they roll down the mountain.

Dialogue can be another problem – if your character speaks, how do you make it sound as if it's coming from the player? One solution has been to process the recordings using impulse responses of a human skull!

As you work through this book, we'll encounter other areas and aspects unique to game audio, but for now, let's jump to Chapter 4, where we'll take our first proper look at FMOD.

Getting Started with FMOD

4

Learning Outcomes

By the end of this chapter, you will be able to:

- Navigate the FMOD user interface.
- Use the FMOD Transport controls.
- Identify the different FMOD Instrument types.
- Create a new FMOD project.
- Create a new FMOD Event.
- Create and configure Single and Multi Instruments in FMOD.
- Assign FMOD Events to Banks.

FMOD Studio has been designed to look familiar to anyone who's used a DAW, but, as we'll see, it has some very different capabilities.

When you open up FMOD Studio, you are presented with the options to check out the latest updates, access the manual, or open a recent Project.

FMOD comes with a demo project. We'll be using this while we take a look around the interface, so select Examples.fspro. This will take us to the Event Editor. This is the window we use to configure our Events, and we'll be spending most of our time in FMOD here.

Figure 4.1 FMOD overview

On the left-hand side, you can see that there are three Tabs available: Events, Banks and Assets (you can show and hide this section with the [B] key). Stay in the Events Tab.

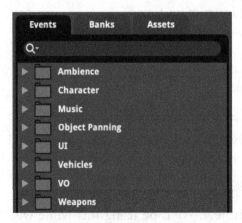

Figure 4.2 FMOD Events, Banks, and Assets Tabs

The Events are organized into folders. It will be worth exploring all of these in your own time to see what type of Events are possible, but for now, click on the disclosure arrow for the UI folder, and select the "Okay" Event.

As we mentioned in Chapter 3, an Event is triggered by the Game Engine. The "Okay" Event would be triggered by the player confirming an action in the game, and would play a sound effect. Try it out – hit the Play button at the top of the Event Editor window (or hit the spacebar) to hear the sound.

That's about as simple as an Event can get, so we'll stay here while we take a further look at the FMOD interface.

Transport

The Stop, Play, and Pause buttons are fairly self-explanatory, but you can trigger these from your keyboard:

- Play/Stop: [spacebar]
- Pause: [shift]+[spacebar]
- Re-trigger Event: [return] or [enter]

21

Figure 4.3 FMOD Transport

FMOD also has a built-in feature labeling system – simply hold the mouse over a button, wait a few seconds, and a floating note appears.

When the Event is triggered, you see the Playback Position moves through the Timeline. As we'll see in later chapters, the Timeline is actually a special kind of Parameter, that increases its value over time. We'll be treating it in much the same way as the time ruler on a DAW.

Just to the right of the Transport buttons, we have the option to display the Timeline value as Time (Minutes:Seconds:Milliseconds), or Beats (Bars:Beats:Sub-beats).

We'll be using Beats for our music in later chapters (where we'll also have to define the Event tempo), but for now, ensure it's set to Time.

Then we have the Loop Playback option. This is only used when auditioning Events – it will have no effect on the eventual sound in the game. Turn this on, hit play, and you'll hear the Timeline loop.

Under this is the Follow Playback Position – but we'll come back to this in just a moment.

Finally, we have the Tracks and Strips view options. Change over to Strips view, and you'll see a fader that we can use when mixing Events. We'll take another look at this later on, so put it back to Tracks view.

The Timeline

Under the Transport is the Timeline Tab. Every Event has a Timeline.

The "Okay" Event Timeline has one Audio Track, also named "Okay." Audio Tracks do not directly play audio files. Instead, they contain "Instruments." There are several types of Instrument, which can cause different things to happen:

- Single Instrument: contains a single audio file.
- Multi Instrument: can contain multiple audio files.
- Event Instrument: can contain other Events (for example, one Event can be nested inside another Event).
- Scatterer Instrument: can contain multiple audio files, and continuously triggers playback instances. Useful for creating ambiences (see Chapter 16).
- Programmer Instrument: can play any audio file – even from outside the FMOD Project. These can be useful when localizing dialogue (we won't be using these for our game).
- Plugin Instrument: for custom-developed Instruments. For example, FMOD comes with trial plugins of AudioGaming's AudioWeather and AudioMotors. As the names suggest, these are used to create weather and engine sounds (these trial versions cannot be included in a game build).
- Snapshot Instrument: can trigger changes to the audio mix.

Localizing

Localizing is where changes are made to a game so that it can be sold to different markets. This usually involves changing the language – for example, in-game dialogue, user interface, and graphics probably need to be altered.

Localization can be a much greater challenge than you might first think – changes may need to be made to suit the differences in language, culture, and market.

Trigger Regions

It's important that you don't just think of Timelines, Instruments, and the Playback Position in the same way as a DAW timeline.

An Instrument has a "Trigger Region," indicated by the colored boxes in the Event Editor. You can see one in the "Okay" Event – it's the blue box labeled "UI1okay," with an audio waveform displayed on top.

When the Playback Position is over the Trigger Region, the Instrument will play.

OK, for our current Event, there's no obvious difference between this and a DAW, but in later chapters, we'll see that we can set up very different behaviors.

As well as our Audio Track, we have a Master Track. We can add additional Audio Tracks to an Event, so this allows overall control of the Event output signal.

To zoom in and out of the Timeline, hold [opt] (Mac) or [alt] (Windows) and use the scroll wheel. Alternatively, you can use the square-bracket keys ([ ) and ( ]). Experiment with this, and you'll see that there's a limit to how far you can zoom out. This is determined by the length of items in the Timeline.

Once you've zoomed in enough so that the entire Trigger Region is no longer visible on the screen, you can try out the Follow Playback Position setting in the Transport. With this turned on, your view follows the Playback cursor. Turn it off, and the view remains fixed in the Timeline.

To change track heights, either move your mouse cursor to the boundary between tracks until it changes to a double-ended arrow, then click and drag (or use the [cmd]+[[] (Mac) or [ctrl]+[[] (Windows) shortcut).

23

Birdseye View

The Birdseye is an overview of the entire Event (if you're a Pro Tools user, you'll recognize this as being very much the same as the Universe).

With such a simple Event as this, you're unlikely to need the Birdseye. However, try zooming out, and you'll notice a gray box. This illustrates the current Event Edit view, and you can drag this (or click on the Birdseye) to navigate around. This can be a real timesaver in complex layered Events. It's also possible to zoom in and out by dragging the edges of the gray box (but I tend to find this quite fiddly and awkward).

The Deck

The Deck is located just under the Birdseye, and it changes according to whatever you have selected. You can show and hide this using the [D] key.

Select the UI1Okay Trigger Region, and the Deck will change to show you the audio waveform and parameter settings for the Single Instrument. A yellow box surrounds the Trigger Region, indicating that it has been selected.

Now select the Audio Track (click anywhere on the gray-space around the track name). You'll see the yellow box is now around the Audio Track and the Deck will show you the Track controls.

Finally, select the Master Track. As you'd expect, the yellow box is drawn around the Master Track, and we can see the Master Track controls in the Deck.

The Overview

Over on the right-hand side, we can see the Overview. This allows us to add Tags and Notes to the Event. However, on 3D Events, the Overview also shows us a 3D preview.

Change to the Weapons > Full Auto Loop Event. Check your monitor levels (it might be loud!), and hit Play. The sound effect will loop, and you can use the 3D Preview in the Overview to see how the sound will behave in 3D space. You can move the sound source further away from you, and pan it around (note – unless you have a surround sound monitoring setup, you won't be able to hear it move behind you).

Explore the Examples FMOD project for yourself. Experiment with the different Event and Instrument types – see if you can figure out what they do. Then come back, and we'll create our own FMOD Project.

Creating Your FMOD Project

It's possible to have multiple FMOD projects open at once, so if you haven't done so already, start by closing down the Examples project.

- Go to File > New.
- Go to File > Save.
- Name your project. I tend to use the name format FMOD_ GameName, so this will be FMOD_MyFPSGame.
- Take this opportunity to create a folder to contain both your FMOD and Unity projects. Name this "MyFPSGame," and save the FMOD project inside.

Note: I have a dedicated drive on my computer for work in progress, but it's entirely up to you where you store your work (however, avoid using a USB memory stick, as they tend to be slow and unreliable).

If you navigate to where you saved your work, you'll see that we've actually created a number of files:

- A folder.
- A file with the extension .fspro (this is the FMOD Studio file extension).
- A sub-folder called "Metadata," which contains more subfolders.
- Many.xml files.

We're not going to be accessing these folders and files directly, but they're needed by the FMOD project. This means that when you're backing up or transferring your work, you need to ensure all of these files are included. If you were to only include the .fspro file, you'll have left behind nearly all the data – so make sure that you back up the entire folder.

Underscore, camelCase, and PascalCase

You should avoid using spaces in your file names, as it can cause problems when coding (you may have seen spaces in file names represented as "%20" before).

Instead, you can use underscores (for example, ciaran_robinson), or camel casing. This is where capital letters are used to separate words (for example, "camelCase").

There is a subset of camel casing called Pascal case, where the first letter is always capitalized (for example, "PascalCase").

We will be using camel casing in our coding, and as we'll see, C# is a case-sensitive language, so it will be essential to copy all code exactly.

Head back to FMOD, and it's time to tell you the secret to using this software: Right-click.

The FMOD interface is designed so that almost every feature you need to access is available via right-clicking. If you're not sure how to do something, move the mouse cursor over to what you think would be the relevant area on the screen, and right-click. Odds are that that's how to do it.

Our first Events will be Footsteps, and Jump and Land vocalizations. However, an FMOD project can get very complex, so before we create these, we should set up our file organization.

We'll need a folder for all our Player Events. Mouse over the Events Tab, and – you guessed it – right-click. Choose "New Folder," and name the folder "Player."

We'll also need subfolders for the Footsteps and Vocal Events, so right-click on the Player folder, and repeat the process. Name these folders "Footsteps" and "Vocals."

Figure 4.4 FMOD Event folder hierarchy

We will be writing code to access the folders, so it is extremely important that your Folder and Event names match exactly – for example, if you name the folder "Vocal" instead of "Vocals," then it won't work. To rename a Folder or Event, double-click (or press [F2]), and enter the correct name.

If you accidentally place one of these files or folders in the wrong place in the hierarchy, simply drag them over to where they need to be (or you can assign them via the right-click menu. This can be very useful when an FMOD project gets too big for all the folders to fit in the Events Tab without scrolling).

Now right-click on the Vocals folder, and choose New Event > 2D Event (we'll be looking at 3D Events in Chapter 6). Name this Event "Land."

This gives us an Audio Track in the Timeline. It's called "Audio 1," so double-click on the name to call it something more descriptive. I'll call it "Vox," but as the track name is only seen inside FMOD (it won't be accessed by Unity), you can name it anything you like.

Now we need to bring in our audio files. There are several ways to do this, but let's start with the long way round. This way you'll see exactly what's happening, before I show you the quicker methods.

Go to File > Import Audio Files. Navigate to your samples folder. I'm going to be using the samples that I created for this book, but if you want to use your own sound effects, that's great too.

Select one of the VO_Avatar_Land files, and click on "Open."

This opens up the Audio Bin, which shows you all the audio files imported into the project. From here you can audition the files, apply data compression, and set audio to stream from disk (see Chapter 28). You can access this window at any time via the Window > Audio Bin menu.

Now drag the audio asset onto the Event Timeline. This has created a new Single Instrument, and set the Trigger Region to match the duration of the audio recording.

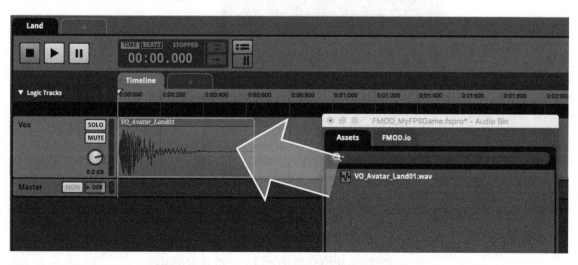

Figure 4.5 Jump Single Instrument

You're probably too far zoomed out of the Timeline to clearly see where you've placed the Instrument, so zoom in and drag the Trigger Region all the way to the start.

Try it out, and hopefully you'll hear the vocal sample play. It should sound okay, but try playing the sound several times…

It doesn't take long before hearing the same sample play again and again becomes irritating. While repetition might be fine for some Events (for example, a UI sound effect), it's unsuitable for others. Imagine Skyrim-hopping up a mountain, and having to hear the same land sound every time…

One solution to this is to use Multi Instruments. These can hold multiple audio files, and, by default, will randomly play a different one every time the Instrument is triggered.

Select the Single Instrument you just created, and delete it (you can use [Delete] or the right-click menu).

Go back to File > Import Audio Files. Select the remaining VO_ Avatar_Land files (you don't need to re-import the existing file), and click on "Open."

This brings us back to the Audio Bin. Before we go any further, let's get this organized. Select all of the Assets, right-click, and choose "Move Into New Folder." Name this "VO_Avatar_Land."

Figure 4.6 Audio Bin organizing

Selecting Assets

FMOD allows you to select assets in the same way as a computer browser.

- Click on a different asset to change selection.
- Click in the gray-space to de-select all assets.

- [cmd]-click (Mac) / [ctrl]-click (Windows) to add (or remove) non-adjacent assets to your selection.
- [shift]-click a second asset to add an asset, and all items between, to your selection.

We now have two options:

First, you can right-click on the Audio Track, and choose "Add Multi Instrument." In the Deck, you'll see the Playlist, which tells you "Drop sounds here." There's no prizes for guessing that you can then drag the audio assets from the Audio Bin into the Playlist.

While this works, you would then need to adjust the Trigger Region so that it is long enough to cover the longest audio file.

Instead, don't create the Multi Instrument via right-clicking. Simply select all of the audio assets, and drag them straight onto the Audio Track. This creates a Multi Instrument, and automatically sets the Trigger Region appropriately (there's actually another way around this via the Async parameter, but we'll look at this in Chapter 5).

Finally, it's actually possible to drag audio files straight from your computer file browser onto the Audio Tracks. While this is a quick and easy option, you should still take the time to go to the Audio Bin and ensure that everything is organized – as your projects get bigger, organization becomes more and more important.

You can see that the Multi Instrument shows all of the audio file waveforms.

Try playing the Event several times. You should hear that a different one plays each time you press play (if you're hearing the same sound every time, check that the Randomize Playlist option is selected – it has the icon of a die).

29

Randomize vs Sequential

If you turn off the Randomize Playlist option, then you will hear the first audio file play every time the Event is triggered.

If the Timeline is looped, then you will hear the audio files play in the Playlist order.

This has changed from earlier versions of FMOD – previously, each time the Event was triggered, the audio files in the playlist would have been played sequentially.

Figure 4.7 Multi Instrument

You'll probably have noticed that one of the Land vocalizations sounds somewhat out of place: VO_Avatar_Land_Ow01.

I'd fully understand if you decided to remove this from the Multi Instrument. However, I'm going to use this to demonstrate Probability Weighting, by setting the vocalization so that it only plays one time in every 100.

An Easter Egg is a secret item in a game. The first documented one of these is in the game *Adventure* on the Atari 2600. If the player wandered round long enough, they'd find a one-pixel dot. If this dot was carried to another area in the game, it gave you access to a secret room, where you could see the message "Created by Warren Robinett."

You'll still find Easter Eggs – rare and hidden events and items that reward the player for putting time into a game. A great example of these are the reload animations hidden in the *Battlefield* games. These are triggered only one time in 10,000, and range from simply spinning a pistol, to a third arm appearing to the side and passing you an ammunition clip. I'm going to use the "ow" vocalization in a similar way.

Select the Multi Instrument, and in the Deck, right-click on the VO_Avatar_Land_05 file, and choose "Set Play Percentage."

At the moment, you can see that it's set to 20% – the play percentage is split equally between all the files.

Double-click on the value, and enter "1." That's all it takes. Try playing the Event a few times, and it's quite unlikely that you'll hear the "ow."

Figure 4.8 Multi Instrument probability weighting

Now create a second 2D Event in the Vocals folder. Name this "Jump," and use the VO_Avatar_Jump files in a Multi Instrument.

Next, create a new 2D Event in the Footsteps folder. Name this "FootstepsGround" (later, we'll be creating a second Event for footsteps on a metal surface).

Rename the Audio Track as "Feet," and use the FS_Ground_Boot_Walk files in a Multi Instrument.

31

Recording Footsteps

Footsteps fall under the category of "Foley" – sounds of people moving (though the term is often used for almost all added sounds). For film and TV, this is performed by a Foley "artist" or "walker." This involves watching the film, and matching the pace and gait of the actors to reproduce and capture the sounds which weren't possible to record on set.

For game sound, the sound effects are triggered by the Game Engine, so there is no performance to match. However, the technical considerations are much the same. You should try to capture clean, dry footsteps that match the weight of the game avatar.

For an FPS game, we know the position of the listener relative to the feet – it will be about around 6 foot (1.8 meters) from the avatar's ears to the ground. Ideally, you should place your mic at a similar position (though you shouldn't necessarily place the mic near the performer's head, as you'll capture too much breathing etc.).

If you're recording your own sound effects at home, you might not have access to a quiet, dry environment, so you'll have to close-mic. You might then need to EQ and process the recordings to match the visuals. However, why not take the opportunity to experiment with recording locations? Get hold of a portable recorder, and capture some foley in the field. Take everything outside, and record yourself walking on grass, concrete, metal – anything you can find. Build yourself a library of footsteps.

If you listen to yourself walking, you'll notice that you don't just hear the sound of your feet on the ground. You can hear the movement of your clothing – and if you walked around with a rocket launcher, I'm sure you'd also hear metallic clinks.

It would be possible to capture these sounds at the same time as the footsteps, but we'll have far more options available if we keep them separate.

Right-click on the existing Audio Track and choose "Add Audio Track." Name the new Track "Cloth," and use the FOL_Cloth_Walk_Bag audio files in a Multi Instrument.

When you listen to this back, you'll notice that the Cloth sound effects mask out the footsteps, so turn down the level of the Audio Track until you get a suitable balance. It might also help to move the Cloth Trigger Region a little later in the timeline – after all, you should hear the sounds of your legs in between footsteps.

Keeping the Cloth sound effects separate from the Footsteps has had another advantage – it has increased the randomization possibilities of the sound effect. Before, we only had four possible footsteps. Now, each footstep can be played with one of four Cloth sounds, giving us 16 combinations.

Now add a third Audio Track, and name this "Weapon." Use the FOL_Weapon_Movement_RL files in a Multi Instrument.

However, I don't want the Weapon sound to trigger every single step.

Select the Weapon Multi Instrument. To the left-hand side of the Deck you'll see a disclosure triangle labeled "Trigger Behavior." Click here to open it up.

We'll be setting Trigger Conditions etc. in some of our later Events, but for now, click on the icon of the die above the Probability pot.

Now experiment with the Probability setting until you have something that sounds right. I'll set it to 20%, so the Instrument will only trigger one time in every five steps.

Figure 4.9 Footsteps Event

File Names

A game project is likely to have hundreds (if not thousands) of samples and recordings. It is therefore really important to make sure that they are labeled and organized. There are several naming systems, but I tend to use category-based naming, with the format: Category, Noun, Verb, Detail, Number. You can see that this is what I've used for my footsteps:

  FS_Ground_Walk_Boot_01.wav

- They come under the category FS (for Footsteps).
- The steps were on the ground, and they were recorded at a walking pace.
- The type of shoe was a boot.

This lets me know exactly what's in a file, purely from its name.

Once your projects get big enough, you're going to have to invest in SFX Management software – for example, Basehead or Soundminer. These allow you to embed tags in the file's metadata, which can then be used to search and organize your effects, even when they're spread across multiple locations.

You don't have to use the same naming system that I've covered here – but you're going to have to use something!

Banks

Our Events are pretty much good to go. However, we need to "Build" our "Banks."

A Bank is a collection of Events that would be loaded and unloaded into the game memory in one go. For example, you could place all of the Events associated with one enemy into one Bank. The Bank would only be loaded when the enemy spawns into the game. Once it's been defeated, the Bank can be unloaded, freeing up memory.

Our game is going to be quite simple. However, it's never too early to get into good habits (and, as we'll see when it comes to optimizing the game, it can help to keep the Master Bank clean).

The Events Tab shows me that none of my Events are yet assigned to a Bank, so select all three (you'll need to command-click (control-click on Windows) to select non-adjacent items) and right-click. Choose Assign to Bank > New Bank, and name this "PlayerBank."

Although we've assigned the Events to a Bank, we haven't yet "Built" the Bank. These are the files that we need to export from FMOD, which can be used by our Unity project.

Save the FMOD project (File > Save). Then go back to the File menu, and choose "Build" (you can also use [F7]).

Before we move on to Chapter 5, let's take a quick look at the other two Tabs: Banks and Assets.

The Banks Tab

In the Banks Tab, you can see we have a Master Bank, as well as the PlayerBank we created. If you open the disclosure arrow, you can see our three Events in the PlayerBank.

If you need to move an Event from one Bank to another, it's simple enough to drag and drop between them.

The Assets Tab

This is the same Tab we saw in the Audio Bin.

In Chapter 5, we'll be taking a look at looping in FMOD, as well as seeing how to prepare loops for use in a game. Before jumping ahead, why not take the time to fine-tune your existing Events. How about adding an impact sound effect to the Land Event? Or recording your own vocalizations and footsteps? The more you customize your game, the more fun you'll have making it.

5 Looping

Learning Outcomes

By the end of this chapter, you will be able to:

- Edit audio files in a DAW so that they are suitable for use as looping assets in a game.
- Create Loop Regions in the FMOD Logic Track.
- Configure FMOD Instruments as Asynchronous or Timelocked.

In this chapter, we'll look at the various types of looping available in FMOD – but before we do this, we'll need to take a look at how to prepare our audio files so that they loop correctly.

FMOD has very limited audio editing capabilities, so you'll need access to a DAW. I'm going to be using Reaper, but the principles covered can be applied to any DAW.

Reaper

In film and TV post production, Pro Tools is the leading DAW. There isn't yet a standard DAW for game audio (though Sternberg's Nuendo is making headway with its integration with Wwise).

In recent years, Reaper has been adopted by many game audio professionals. This is no doubt partially due to the price (a personal license starts at $60), but also because of its scripting capabilities. These allow you to create macros and batch processes, which can be invaluable in a development workflow.

Reaper offer a free 60-day evaluation license, and I thoroughly recommend trying it out.

If you take an audio file and loop it, there's a very good chance that you'll hear a click at the loop transition. Why does this happen?

Well, it's due to the sudden change in waveform level. If you zoom in and look at both ends, you'll see that they are at different sample values.

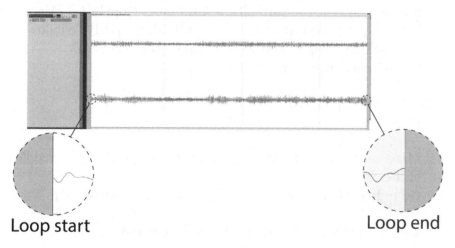

Loop start   **Loop end**

Figure 5.1 Waveform sample jump

This sudden jump from one level to the other creates the click.

In a DAW, you can remove the clicks by cross-fading over the loop. However, this isn't possible in FMOD, so let's see how to get around this.

Reaper tries to avoid these clicks by automatically adding fades when media is imported or edited. This is usually helpful, but here it's going to cause us problems, and it will be better to fix them manually.

• Go to Options > Preferences.
• In the Preferences menu, go to Project > Media Item Defaults.

- De-select the "Create automatic fade-in/fade-out for new items, length" checkbox.
- Click on Apply.

Next, open your DAW, and import the file "SFX_SentryDrone_Humm_NotLooped_01.wav" (in Reaper, you can do this by going to Insert > Media File). We'll need a cleanly looping version of this in the next few chapters.

Now trim off the ends of the file. This will give you "handles" for cross-fading later on.

In Reaper:

- Select the SFX_SentryDrone_Humm_NotLooped_01 clip.
- Click and drag to select the section you wish to keep (you might find this easier if you turn off the Grid Lines ([opt]+[G] (Mac) / [alt]+[G] (Windows)), or to hold [cmd] (Mac) / [ctrl] (Windows) to disable the snap to grid while making your selection).
- Go to Item > Trim items to selected area.

Now slice the clip in two.

In Reaper:

- Make an "Insertion Point" (click without dragging) in the middle of the clip.
- Go to Item > Split items at cursor (or use the shortcut [S]).

Now we need to swap the two halves of the clip around.

In Reaper:

- Make sure that Snap is enabled ([opt]+[S] (Mac) / [alt]+[S] (Windows)).
- Click and drag the left half of the clip over the right, swapping their order.
- Select both clips, right-click in the timeline ruler and choose "Set selection to items."
- Ensure that "Repeat" is turned on (from the Transport controls, or with the [R] shortcut).

Now try playing the loop, and you'll notice that it will now transition smoothly – however, there is still probably an audible click in the middle.

Why is there no click over the loop transition? Well, the end of the loop started off next to the beginning of the loop, so there is definitely no sudden change in sample level.

But what about the click in the middle? Well, we're going to use a cross-fade to fix this.

- Select both clips.
- Make a selection over the splice point.
- Press [X] to create your cross-fade.
- Select both clips, right-click in the timeline ruler and choose "Set selection to items" to re-select and audition your loop.

It might take some experimentation with the cross-fade settings to get this right (double-click on the cross-fade to access the editor), but you should be able to achieve a smooth transition.

All we need to do now is to export our loop.

39

- Select both clips.
- Go to Items > "Glue items."
- Go to File > Render.
- Set the Source to "Selected media items."
- De-select the "Tail" checkbox.
- Set the Directory (destination of the exported file).
- Set the File name to SFX_SentryDrone_Humm_Looped_02.wav
- Click on "Render 1 file."

You now have a smoothly looping file that will not require the use of cross-fades.

When creating music loops, there is another issue that needs to be addressed. If you play a four-bar loop in your DAW, it should loop smoothly. However, if you were to bounce this down and import it to FMOD, the loop transition is likely to sound far more obvious.

This is because when looping in the DAW, any reverb tails or Instrument note releases are not choked when the Transport returns to the start.

When bouncing down the file, there will be no reverb tails etc. at the beginning, and as a result, they will be abruptly truncated every time the loop restarts. We need to capture these tails, and place them at the start of the loop.

To fix this in Reaper:

- Right-click on the track (not the clip) and select Render/freeze tracks > Render tracks to stereo stem tracks (and mute originals).
- Select the original clip.
- Right-click in the timeline ruler and choose "Set selection to items."
- Go to Item > Split Items at time selection ([shift]+[S]).
- Select the Rendered track, and go to Insert > Track.
- Drag the reverb tail into the new track at the start of the Timeline.
- Go to File > Render.
- Set the Source to "Master mix."
- Set the Bounds to "Time selection."
- De-select the "Tail" checkbox.
- Set the Directory and name the file.
- Click on "Render 1 file."

This technique should be used whenever creating music loops – whether it's for use in a game, or if you're making beats (if you're using Apple's Logic, then this can be done automatically by selecting the "Bounce 2nd Cycle Pass" option from the bounce menu).

Now you can create loops that will work in FMOD Events, let's see how we can use them. Our game is going to have Sentry Drones, which will scan for the player before opening fire. These will make a humming sound which will need to make use of looping.

- In FMOD, create a new Event folder, and name this "Enemy."
- Inside the new folder, create a new 3D Event (we'll be covering these in Chapter 6), and name this "SentryDroneNoises."
- Use the SFX_SentryDrone_Humm_Looped01.wav file in a Single Instrument in the DroneSound Event.

There are several types of looping in FMOD. Let's start by looping the Instrument.

Instrument Looping

Select the Single Instrument, so that we can see its settings in the Deck.

To the top right of the waveform you can see a loop icon (and a loop count value, which is grayed out and indicates "1"). Leave this turned off for now.

Place your mouse cursor at the right-hand edge of the Trigger Region, and you'll notice it changes to an icon of two arrows. Click and drag, extending the Trigger Region to the right.

You'll now see a flat line displayed after the waveform in the Trigger Region. Try playing the Event, and, unsurprisingly, once the Playback Position moves beyond the waveform, the sound stops.

Now go back to the Deck, and turn on the loop icon (it will change color to yellow). The waveform in the Trigger Region changes to show that the file will now loop. The loop count value also changes to show ∞ (infinite), and you can also click in here to set how many times it will repeat.

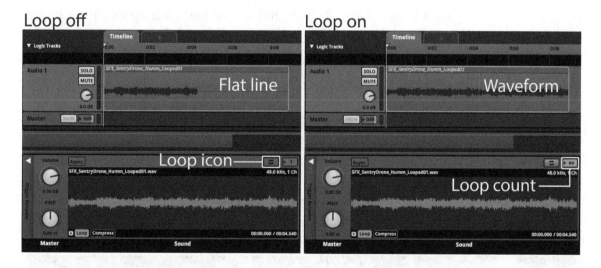

Figure 5.2 Looping in FMOD Events

That will work for some situations, but this Event will be used to add a sound effect to the Drones in the game. We can't simply extend the Trigger Region, as we don't know how long they will need to play for. Instead, we'll need to loop the Timeline. This will involve using Logic Tracks.

Logic Tracks

The Logic Tracks are located just above the Audio Tracks in the Timeline. There's a disclosure arrow that allows you to show and hide them.

By default, there's nothing in the Logic Track, but try right-clicking here, and you'll see that there are several options available:

- Add Destination Marker.
- Add Tempo Marker.
- Add Loop Region.
- Add Sustain Point.
- Add Transition.
- Add Transition Regions.

42

Try adding a Loop Region, and a blue line with arrows at each end appears in the Logic Track. You can click and drag these arrows to set the duration and position.

However, there's an easier way to ensure that the Loop Region is placed correctly.

- Right-click on the blue line, and choose "Delete."
- Right-click on the Trigger Region, and choose "New Loop Region."

This will create a new Loop Region, placed exactly over the Trigger Region.

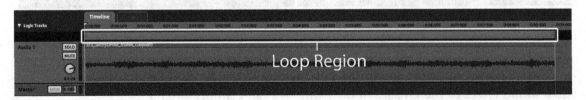

Figure 5.3 Loop Region labeled

Async

If you've left the Trigger Region extended, when you play the Event, the looping will sound wrong. After we went to great effort to edit our files so that they looped smoothly, we're now looping from the middle of the audio file.

Back in the Deck, there's an option for "Async." Click on this, and the Trigger Region changes to show the original waveform, but stretched out over the entire Region duration.

The Instrument is now "Asynchronous." This means that it has its own independent playback. Play the Event, and you'll see it has its own playback cursor.

This will suit our looping Instrument. As long as the Playback Position is over the Trigger Region, the sound will loop correctly.

Of course, we could have set our Loop Region to precisely cover one loop, but if you are using loops in multiple layered Audio Tracks, the Asynchronous option means all of the files don't need to have the exact same duration.

Asynchronous (as opposed to "Synchronous" or "Timelocked") Instruments are not just useful when looping. They are ideal for when you need to have a sound play in its entirety once triggered – for example, in Chapter 4, rather than ensuring that the Trigger Regions were long enough to cover the longest vocal and footsteps files, we could have simply set the Instruments to Asynchronous. You might want to go back and change these…

Synchronous Instruments still have their uses – for example, when configuring interactive music. You can jump playback into the middle of a Synchronous Instrument. This isn't possible with an Asynchronous Instrument, which will always play from the beginning of the audio file. You can also trim the start and end points of a Synchronous Instrument by editing its Trigger Region.

43

Cut

You might have noticed that a button labeled "Cut" appears next to the Async button when it's pressed. This determines what happens when the Playback Position is no longer over the Trigger Region.

- By default, an Asynchronous Instrument will complete playback of its sound after the Playback Position leaves the Trigger Region.
- If "Cut" is enabled, the Instrument will immediately stop the sound the moment the Playback Position is no longer over the Trigger Region.

In Chapter 6, we'll be using the SentryDroneNoises Event to look at the differences between 2D and 3D sound, so before you move on, make sure that your Instrument is looping correctly.

44

2D vs 3D Sound

6

Learning Outcomes

By the end of this chapter, you will be able to:

- Define the differences between 2D, 3D, stereo, and mono.
- Use the FMOD Spatializer parameters to control how an Event behaves in a 3D game space.
- Use the FMOD Sandbox to test how an Event behaves in a 3D game space.

Our Events can be 2D or 3D – but don't get this confused with surround sound.

A 3D Event is one that can be localized. It has an origin point in the game world. When the player moves closer towards the sound source, it gets louder. As the player turns towards the origin, the sound will pan towards the center.

A 2D Event does not change in position as the player moves around the game.

Most diegetic Events will be 3D. In our game, we will be using 3D Events including:

- Sentry Drone sounds
- Laser firing sound effects
- Explosions
- Rocket trails
- Ambience
- Crate Collision sound effects

Most non-diegetic Events (for example, the game music and UI sound effects) will be 2D.

If an Event is 2D, that does not mean that it can't have stereo properties – it can still be panned. Think of this as like wearing headphones when listening to music. You can still listen in stereo, but turning your head and walking around the room doesn't change the position of the Instruments in the mix.

Mono vs Stereo

Stereo

"Stereo" or "stereophonic" is one of the most misunderstood terms in audio. It does not mean "two speakers." A stereo signal is one that contains localization information (in fact, the word "stereo" comes from the Greek for "solid," meaning that it has a position in 3D space).

If I click my fingers, this produces a stereo signal – I can tell where the sound is coming from. However, if I use a single microphone, although I can record a finger click, I cannot capture the positional information. To do this, I would need to use at least two microphones.

Using more than one microphone doesn't mean that your recording is stereo either – they have to be configured in a "stereo array" – otherwise, you just get a dual monophonic recording.

Mono vs Stereo Files

When files are exported from a DAW, care needs to be taken to ensure that they are bounced down in mono or stereo appropriately. To optimize our game, file sizes should be kept as small as possible, and a mono audio file would be twice the size if it was two-channel stereo.

(But don't think that you can't use monophonic files – when they are used in the game, they are given a virtual origin [i.e. will be heard in stereo]).

If a sound is panned to the middle, then there is no difference between the left and right channels, so it should be exported in mono.

Consider how the sound will be used in the game. Sound originating from a single point source does not have a stereo width, so mono files should be used.

Sound sources with a stereo width (e.g. ambiences and atmos sounds) will need to use stereo files.

This issue becomes more complicated if reverb is pre-rendered on sound effects (as opposed to added "live" by the Audio Engine). Reverb sounds odd in mono, and should not come from a single point origin.

There's no simple answer to this, apart from try it and see how it sounds.

Ambisonics

Ambisonics is a 3D surround sound system that allows for sound to be positioned around, above, and below the listener. Unlike other surround sound formats, it does not use separate channels for each speaker. Instead, it is matrixed into "B-format," which can then be decoded into whatever channel format is necessary for the playback system.

This fell out of use after its development in the 1970s, but it's recently had quite a resurgence, as it's ideal for use in game audio (especially for virtual reality). If you want to experiment with this, check out the Facebook 360 Spatial Workstation. This is a free set of plugins (VST and AAX) that allow you to design and create 3D audio files (fantastic for atmos tracks).

47

It's not as straightforward as just making all diegetic sounds 3D. For example, we've set our Jump, Land, and FootstepsGround Events as 2D.

It would be possible to set these Events to 3D, and to give them an origin point coming from the game avatar. However, this is unnecessary, as they will not change their relative position to the listener – after all, your feet keep the same distance from your ears as you walk (we'd need to change this for a multiplayer game, as we'd have multiple listening positions to consider).

Take a look at our SentryDroneNoises Event. We've already seen how 3D Events have a 3D Preview in the Overview, so we can see how it will sound in the Game world.

Now select the Master Track, and in the Deck, you can see the Spatializer.

This is essentially the only difference between 2D and 3D Events – in fact, you can convert one to the other by simply adding/deleting the Spatializer (click on the plus symbol, or right-click in the Deck, and choose Add Effect > FMOD Spatializer).

Spatializer Controls

Min and Max Distance

Technically, this value is in "Game Units," but in Unity, 1 Game Unit is equal to 1 meter.

Once the sound source is closer than the Minimum Distance, it won't get any louder. This can be useful for environment ambiences – you may be able to hear the ambience of a room from outside, but once you enter the room, the sound won't appear to come from a specific location (it's also useful to prevent you from deafening the player if they get too close to the sound source!).

The Maximum Distance is the distance of the player from the sound source where it doesn't make a difference if you go any further away. This will usually be the point at which you can't hear the sound at all, but it depends on how you set the Distance Attenuation curve...

Distance Attenuation Curve

The default setting is "Linear Squared." This gives a reasonably realistic attenuation over distance, and (most importantly) the sound level is reduced to zero at the Maximum Distance.

Linear Attenuation isn't very realistic, but it's very predictable – moving 1 meter from the sound source will have the same amount of attenuation, regardless of the original distance from the sound source.

Inverse is the most realistic attenuation curve – it attenuates at 6dB for every doubling of distance (according to the inverse square law). However, this isn't often used, as there is no exact point at which the sound becomes silent – a louder sound will be heard at a greater distance than a quiet one… This can make mixing difficult, and has a knock-on effect on the signal processing load.

Inverse Tapered combines realism with practicality – sound attenuates at 6dB per doubling of distance, but after the Maximum Distance, it tapers off to zero.

Finally we have "Off." A sound can still have a panned location in the game world, but it doesn't get any quieter over distance.

Envelopment

This controls the stereo image width of the audio signal over distance – as you move further away from a sound source, its stereo width should get narrower. When this is changed to "User," you have access to two additional parameters ("Off" sets both values to zero):

Min Extent

This sets the stereo width angle for when the listener is at the Max Distance.

49

Sound Size

This sets the virtual size of the sound source – essentially, the maximum stereo width (though of course, this will depend on the listener position).

When this is set to Auto, the Min Extent is set to 0 degrees, and the Sound Size is set to twice the Min Distance.

Pan Override

There's a gray box with a lighter gray disclosure triangle just to the left of the Envelopment controls. Click here to access the Pan Override controls. This allows you to offset the pan position and stereo image spread.

The Mix pot controls the amount of offset. When this is set to 0%, the pan position is only controlled by the sound source's 3D position (i.e. Pan Override has no effect). At 100%, localization is entirely controlled by the Pan Override.

The Sandbox

While it can be useful to experiment with the 3D Preview, this only allows you to hear one Event at a time. When creating a sonic environment, we will want to hear how all of the Events will work together.

To help with this, FMOD Studio has a "Sandbox" to test your Events. This is essentially a very (very!) basic Game Engine.

To place an Event in the Sandbox, we need to assign the Event to a Bank, then Build the Banks.

- Right-click on the SentryDroneNoises Event in the Events Tab, and select Assign to Bank > New Bank.
- Name the bank "SentryBank."
- Save the FMOD Project.
- Go to File > Build.

Now you can go to Window > Sandbox.

The circle with the arrow inside represents the listening position. You can use your mouse scroll wheel to zoom in and out, and you can click and drag the listener around. However, there isn't much point doing this until we put some sounds into the Sandbox, so drag the SentryDrone Event over from the Events Tab.

This adds a circle with several circle sections inside – these indicate that the Event is currently playing, so hopefully you should be able to hear the SentryDrone sound effect. If not, drag the circle closer to the listener. As you move this around, you should hear the position and level of the sound source change.

When the Event is selected, you'll see a gray circle surrounding it – and if you look closely, there's another, fainter circle inside. These represent your Min and Max Distance values.

You should also see a green box at the bottom of the screen that says "Live Update On." This means that you can go back to the Event Editor, change any of the settings, and these changes are immediately effected in the Sandbox.

You can use the Transport controls at the top of the Sandbox to control playback, but there are more options available by right-clicking on the Event circles.

Play, Stop and Pause are self-explanatory (as is "Remove"), but Looping is slightly less obvious. This loops the Event so you don't have to trigger it again and again (this is for testing purposes only – it has no effect on the actual Event).

Add the FootstepsGround Event to the Sandbox, and it will play just once. Right-click to turn on Looping, then right-click again to start the Event playing. The FootstepsGround Event will now repeat until you stop it (or close down the Sandbox).

The Sandbox is a useful way to quickly test out Events, but it has its limitations. For example, the settings are not saved with the FMOD Project, and you can't use it to test out Doppler shift. Not to worry – in Chapter 7 we're going to start looking at Unity, and you'll be trying out your sounds in a true 3D game world very shortly!

7 Getting Started in Unity

Learning Outcomes

By the end of this chapter, you will be able to:

- Create a new Unity project.
- Navigate the Unity interface.
- Organize the Unity project assets.
- Manipulate GameObjects in the Unity scene.
- Navigate the Unity scene.

If you remember from earlier, when you open up Unity, you see the Projects menu.

Click on the option for "New."

Give your project a name – I'll call mine "AudioForGamesFMODandUnity." Leave it set to the 3D default.

The organization name will already be filled in from when you created your Unity account, so all that's left to be done is to set the Save location.

It's possible to bring in Asset Packages from here, but this only allows you to bring in an entire package in one go. There's another way to do this, which gives us the option to choose what parts of an asset package we want to import into our project. We'll do this very shortly, so leave this option alone for now.

Now click on the three dots to browse to where you're planning to save your Project. If you remember from Chapter 4, we created a folder for this called "MyFPSGame," so navigate here and click on "Open."

This brings us back to the Projects menu, so click on "Create Project," and wait for Unity to do its thing...

You should now see something similar to Unity's default layout – the only difference is that on the right-hand side, you have the Services Tab. We're not going to be using this, so we can safely close it down.

You can do this by right-clicking on the Tab title and selecting "Close Tab," or go straight to the Layout setting at the very top right, and choose "Default."

We'll be sticking to the default layout for the majority of this book, so if you ever get lost, you can always reset from here.

Let's have a look around the layout.

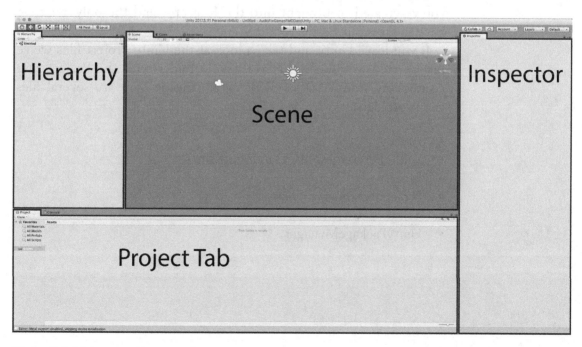

Figure 7.1 Unity overview

The Scene

In the middle of the screen is the Scene. The Scene is where you set out the game world and create your level. Our game will have three of these: where you play the game, as well as the "You Win" and "Game Over" screens.

There are two additional Tabs here – the Game and the Asset Store, but we'll come back to these later.

First, let's save our scene.

- Go to File > Save Scene.
- Give your Scene a name (I'm calling mine "Battleground"), and let Unity save it in its default location.

The Project Tab

You might have just noticed something appear in the tab at the bottom of the screen.

This is the Project Tab. This allows you to access any of the Assets you've created or brought into the Project – and the only asset we have so far is our Battleground Scene.

If you leave Unity, and take a look at the Unity Project files we've just created, you'll see that we have a folder with the Project title (AudioForGamesFMODandUnity), and inside here are several files and subfolders:

- Assets
- AudioForGamesFMODandUnity.sln
- Library
- Project Settings
- Temp
- UnityPackageManager

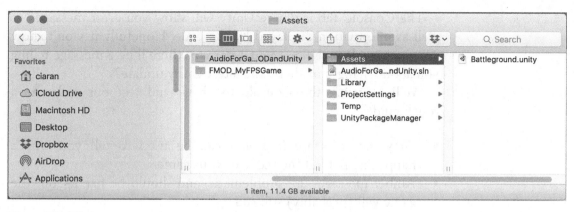

Figure 7.2 File structure

As a general rule, don't touch anything apart from the Assets folder.

If I take a look inside here, I can see a file called "Battleground. unity" – and this is the same folder we were looking at in the Unity Project Tab. In fact, the easiest way to get assets into a Unity project can be to drag the files into this folder (they will then be accessible from the Project Tab).

Let's start as we mean to go on, by keeping our assets organized.

- Head back to Unity, and right-click in the gray-space of the Project Tab. Choose Create > Folder.
- Name the folder "MyScenes" (I tend to name all of my folders using the format "My****," which means that they are all grouped together in the Project Tab).
- Drag the Battleground scene onto the folder to place it inside.
- Double-click on the folder to look inside.

To get back out, you can either use the file navigation menu on the left-hand side of the Project Tab, or click on "Assets" immediately above.

At the bottom right of the Project Tab is a slider. This changes the asset icon sizes – but if you turn it all the way down, you will see your assets in a List view. This can be useful when your Project gets bigger.

The Console Tab

The Project Tab can be switched with the Console Tab – just click on the Tab titles to swap between them.

55

The Console Tab is where Unity will show you error messages, as well as information displayed by the code. Hopefully it won't show you very much at the moment – perhaps there'll be a message telling you about changes made by the latest Unity update?

We'll be using the Console to check and test our code. As a rough guide:

- Gray messages are fine, and can be used to tell you what's happening behind the scene of your game.
- Amber messages are problematic, and should be fixed – but the game will (probably) still run.
- Red messages are bad. The game is probably broken, and the code must be fixed.

We'll come back to this area later, so change back to the Project Tab for now.

The Hierarchy Tab

A GameObject is an item that can be placed at a location within the Scene. The Hierarchy shows you all of the GameObjects that have been used in the game. This is organized into Scenes, so if I click on the disclosure arrow next to the Scene title, you can see that I already have two Objects – a Main Camera, and a Directional Light.

The Hierarchy is often the easiest way to locate and select GameObjects – it can be difficult to click on a specific GameObject in the Scene if it's grouped among many others. You can also use the Hierarchy to find GameObjects in the Scene:

- Select the GameObject in the Hierarchy.
- Move your mouse over the Scene Tab.
- Press [F] (for "find").
- The Scene will zoom and pan to show the selected GameObject.

The Inspector

The Inspector shows you information about the currently selected GameObject. You can think of a GameObject as being a container

56

for Components. It will always have a Transform Component, which allows it to be given a position in the game world.

To give any other property to a GameObject, additional Components must be added. Unity comes with hundreds of Components, and we'll be creating some of our own by writing scripts.

GameObjects

Types are specific forms of data – for example, "integer" (int) is a type that can only be a whole number.

Classes are containers that allows you to group together and manipulate types. "Object" is the base class for Unity. It allows items to be instantiated (instances of them created) or destroyed.

A GameObject is an Object that has a Transform component – i.e. it has a position in the game world.

If you select the Main Camera GameObject, you can see that it has a Transform component, as well as a Camera, a Flare Layer, and an Audio Listener.

There's also an option to "Add Component," and if you have a quick look here, you can see that these are all organized into subfolders.

Feel free to experiment with this – we'll be deleting the Main Camera, so you won't break anything by changing any of the settings.

Navigating the Scene

We don't yet have anything to look at in our Scene, so start by going to Game Object > 3D Object > Cylinder. This will have placed a Cylinder directly in the middle of your Scene window.

In any 3D environment, there are three axes:

- X axis (red)
- Y axis (green)
- Z axis (blue)

You can think of these as X = right/left, Y = up/down and Z = front/back (but of course these will change as your view position moves).

Moving and manipulating 3D objects on a 2D computer monitor is always going to be difficult. Unity uses dedicated tools for this, which you might have already spotted in the top left of the screen.

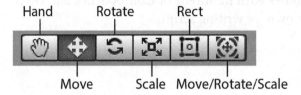

Figure 7.3 Unity Tools

You can select these by clicking with the mouse, or by using the QWERTY keys (note – we're not going to be covering the Rect tool in this book, but it's for manipulating User Interfaces).

Hand Tool (Panning)

This is used for changing our view position. The default Hand tool performs a "Pan", moving our view left and right.

Select the Hand tool [Q], and click-drag in the Scene Tab. This can be achieved using several other techniques, including:

- Hold [opt] + [cmd] (Mac) / [alt] + [ctrl] (Windows) and click-drag in the Scene Tab.
- Use the arrow keys.

After a little experimentation with these, you'll find your own preferred technique. Try using the [Shift] key and panning. What effect does this have?

Hand Tool (Zooming)

We're a little too far from the Cylinder to clearly see what we're doing, so let's zoom in on it. To zoom the view in and out, you can:

- Use the mouse scroll wheel.
- [opt] + [right-click] (Mac) / [alt] + [right-click] (Windows) and drag.

58

While you're zooming, the Hand tool icon changes to a magnifying glass. If you get lost, re-center the Cylinder in the Scene window by selecting it in the Hierarchy, moving your cursor back over the Scene, and pressing [F].

Hand Tool (Orbiting)

So what if you need to look behind an object? You will need to orbit around it. Unity orbits around the center object on the screen.

- Center the Cylinder in the Scene window.
- [opt] + [left-click] (Mac) / [alt] + [left-click] (Windows) and drag.
- You should now be able to orbit the model.

While you're orbiting, the Hand tool icon changes to an eye.

There's one more method that you might want to try out – especially if you're familiar with playing FPS games:

- Hold the right-mouse button down in the Scene Tab.
- Use the mouse to look around.
- Use the WASD keys to move around.

If you're not familiar with this control system:

- W = forward
- S = back
- A = left
- D = right

In addition, Q = down, and E = up.

We can also set our view along the three axes of the Scene. We're going to use the Scene Gizmo for this – you might have noticed this at the top right of the Tab. This acts as a reference or compass to our scene, but the six "arms" also allow us to snap our viewpoint directly along any axis.

Gizmos

Gizmos are visual guides to help configure GameObjects in the Scene. The Tool Gizmos (Move, Rotate, Scale and Rect) are visible automatically when the GameObject is selected. You can configure how the additional Gizmos will appear by setting their checkbox in the Gizmo menu at the top right of the Scene Tab.

Try clicking on the yellow y axis arm. The Scene will snap to an overhead view, with the X axis pointing to the right.

Click on the red X axis arm to change to a side-on view.

The Scene Gizmo also allows us to change between the two view types: perspective and isometric. You can click on the box in the center of the Gizmo (or on the writing underneath) to switch between them.

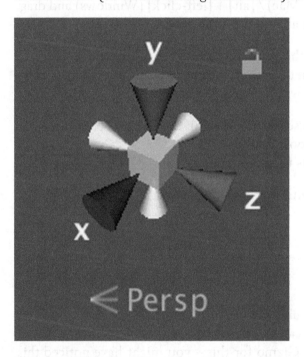

Figure 7.4 Scene Gizmo

Perspective view uses a vanishing point – the lines on the ground look to meet at a point somewhere over the horizon. This appears the most natural looking to the eye. The distance of an object from the viewer will change the position and size.

In Isometric view, these lines remain parallel, and never meet. The size of an object will not change, regardless of the distance from the viewer. This can be very useful when positioning objects in a Scene.

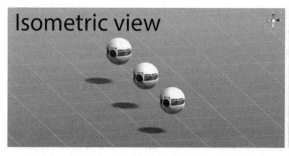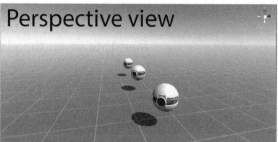

Figure 7.5 Isometric vs perspective views

Switch between the view types, and pan across the screen. The difference between the two should become quite obvious.

The Move Tool

The Move tool allows us to change the position of GameObjects in the Scene.

The placement of an object in a Scene is given by the x, y and z transform position values (x,y,z). 0,0,0 is the center of the Scene, and everything is described as relative to this.

Let's add two more objects to our Scene: a Plane and a Cube.

- Center the Cylinder in the Scene window.
- Select GameObject > 3D Object > Plane.
- Repeat the process to add the Cube (GameObject > 3D Object > Cube).

All three of these objects will now be sitting on top of each other.

Select the Move tool [W] and click on the plane (in either the Hierarchy or the Scene Tabs). We're going to move the plane to the center of the Scene.

- In the Inspector, set the Transform position value to 0,0,0 (tab to go between the values).
- Repeat the process for the Cube.

As you can see, the Cube is half submerged inside the plane. That is because the Transform position value of a cube is taken from the center of the object.

61

The Cube has a size of 1 × 1 × 1 "units" (the default scale for Unity is 1 unit = 1 meter); therefore, we'll need to lift it by 0.5 units if we want it to sit on the plane, so change the Transform position y value to 0.5.

Next we're going to place the Cylinder on top of the Cube. We're going to do this manually.

Ensure the Move tool is selected [W], and select the Cylinder. This time we're going to use the Move Gizmo – the box and arrow that appears when you select an object with the Move tool.

You can move the object in a single axis by using the arrows, or in two axes at once by using the boxes in the center.

- Red arrow: moves the GameObject in the X axis.
- Green arrow: moves the GameObject in the Y axis.
- Blue arrow: moves the GameObject in the Z axis.
- Red box: moves the GameObject in the Y and Z axes (not X).
- Green box: moves the GameObject in the X and Z axes (not Y).
- Blue box: moves the GameObject in the X and Y axes (not Z).

Use these to position the Cylinder on top of the Cube.

You may find that it helps to use the Scene Gizmo to change your view angle. It's worth also experimenting with the parabolic and isometric view settings.

Parent–Child Relationships

We're now going to link the Cylinder and the Cube, so that they can be moved together.

In the Hierarchy Tab, drag the Cylinder onto the Cube. In Unity terms, the Cylinder is now a "child" of the Cube (i.e. the Cube is the "parent" of the Cylinder).

Try moving the Cube in the Scene, and the Cylinder should move with it.

Now select the Cylinder, and examine the Transform position values in the Inspector Tab. You should notice that its position is no longer described relative to the 0,0,0 center of the Scene – it is now relative to the center of the Cube.

The Rotate Tool

This one's fairly self explanatory – we can rotate objects around all three axes.

- Change to the Rotate tool [E].
- Select the Cube (not the Cylinder, though as a child of the Cube it will of course be affected by the rotation).

Click-drag on the the Rotate Gizmo, and experiment with rotating. There are three circles surrounding the Cube:

- Gray sphere: rotates the GameObject in all three axes.
- Red circle: rotates the GameObject around the X axis.
- Green circle: rotates the GameObject around the Y axis.
- Green circle: rotates the GameObject around the Z axis.

If you hold down the [cmd] (Mac) / [ctrl] (Windows) key, you can rotate in 15-degree increments (though it may still be easier to set precise rotation values in the Inspector).

Once you're familiar with the way the tool works, use the Transform Rotation setting in the Inspector Tab to set the rotation to 45,0,0, so that the Cube and Cylinder are leaning forward.

63

Global and Local Tool Handles

We're going to take a moment to look at another of the Transform settings in Unity – the difference between Local and Global. You can see the button to switch between these just to the right of the transform tools – it should say "Global."

Our Cube and Cylinder are rotated at 45 degrees. Select the Cylinder in the Hierarchy (not the Cube).

If you change to the Move tool, dragging the green arrow would lift the Cylinder upwards. However, what if you were trying to recess the Cylinder into the body of the Cube? This would be awkward, and involve carefully adjusting the y and z values.

Leave the Cylinder where it is, and click to change the Tool Handle to Local. Now the X, Y, and Z axes are based on the original position of the Cylinder; before, it was rotated. This will make it much easier to recess it into the Cube. Give it a try.

This parameter becomes grayed out when you select the Scale tool. This is because the Scale tool always works with Local handles, regardless of this setting.

Tool Handle Position

This can be toggled between Center and Pivot, and it becomes important when you're dealing with GameObjects that are nested in a parent–child setup (or if multiple items are selected at once).

When this is set to Center, then the center of rotation (and Gizmo) will be at the middle of all the selected GameObjects – i.e. when you select the parent, the center of rotation will be at the entire parent–child group's center (not the center of the parent).

When it is set to Pivot, then the center of rotation (and Gizmo) will be at the middle of the selected GameObject – i.e. when you select the parent, the center of rotation will be at the parent's center, not the center of the entire parent–child group.

When selecting multiple items at once, the first item selected determines the Pivot position.

Scale Tool [R]

This allows you to change the size of GameObjects in the Scene.

Change to the Scale tool, and select the Cylinder. You can see four cubes that will allow you to change the scaling:

- Gray square: scales the GameObject in all three axes.
- Red circle: scales the GameObject in the X axis.
- Green circle: scales the GameObject in the Y axis.
- Blue circle: scales the GameObject in the Z axis.

However, what if we want to change the size of the Cube? With the current configuration, scaling the Cube will also change the Cylinder.

To scale the Cube independently, we'll need to change our parent–child setup.

- In the Hierarchy, right-click on the Cube, and choose Create Empty. This creates an empty child object at the same location as the Cube (which will eventually be the container for both GameObjects).

- Select the empty GameObject, and in the Inspector name it "MyContainer."
- Now drag it out from the Cube (so that it is no longer a child).
- Finally, drag the Cube and the Cylinder into MyContainer.

You can now use MyContainer to move both GameObjects together, but you can also independently adjust the scale of the Cube.

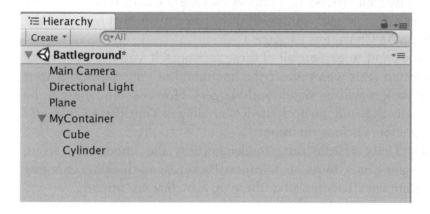

Figure 7.6 My Container

We'll be using empty GameObjects quite frequently – for example, we'll place all of the SentryDrones into an empty GameObject. This will allow us to toggle them between active and inactive at once (it can be difficult to test the game environment if you're constantly being shot at!).

Spend some time experimenting with navigating, adding, and moving GameObjects around the Scene. Once you're ready to move on, select all of the items you've added, then right-click in the Hierarchy to delete them.

Move/Rotate/Scale Tool [Y]

This is a recent, but very welcome addition to the Unity Tools. It gives simultaneous access to all three Transform controls (though you can only Scale in all three axes simultaneously when set to Global Handle Rotation).

In addition, holding the [shift] key changes the Move axis, so that movement is restricted to two axes, relative to your viewing position. Experiment with this, and it will quickly make sense.

Importing Asset Packages

We don't have many assets in our project, but when we installed Unity, we made sure to include the Standard Assets. Now we're going to bring some of these in, which will give us much more to play with.

Start by going to Assets > Import Package > Characters.

After a few seconds, Unity will show you a menu that lists all of the items in the Characters package.

If you want to explore all of these for yourself, you can leave them all selected (this won't affect the final build of your game, but your Unity project will be significantly larger). However, I'm going to be more selective, and uncheck the RollerBall and ThirdPersonCharacter boxes, before clicking on Import.

Give Unity a little time to decompress and import the assets, before going back to Assets > Import Package, and this time choosing "Environment." To bring all of these in, just click on "Import."

Searching Assets

If you now look in the Assets folder in the Project Tab, you'll see that there are two folders – the MyScenes folder we created earlier, as well as a new folder called "StandardAssets."

If you know where Unity places everything, you can navigate through this folder to find what you need. However, there are easier way to do this.

At the top right of the Project Tab is a search menu. Type in a name here, and the Project Tab will show you any items that match. Try typing the word "first," and you'll see a folder, a text document and two scripts that contain "first" in their names.

You can narrow this down by clicking on the Search by Type icon. If you set this to "Script" there should only be two items displayed.

It's also possible to filter assets by labels, which can be useful as a project gets bigger (to add a label to an asset, select it in the Project Tab, and use the Assets Labels option at the bottom of the Inspector).

If you want to keep your search parameters, click on the Star icon. This adds your saved search to the Favorites menu at the left-hand side of the Project Tab (as you can see, there are already saved searches for All Materials, All Models, All Prefabs, and All Scripts).

Figure 7.7 Project Tab

Game Tab

This is where you can see the game. With the default layout, as soon as you hit the Play button, Unity will change to the Game Tab, and switch back when you press Stop.

However, if you change to the Game Tab, you can see an option for "Maximize on Play" at the top right. This will take over the whole of Unity's display with the Game window when you're playing the game. This might be useful later on, but I'd advise you to leave this turned off for now.

If you go to Edit > Project Settings > Quality, you can configure how the game will behave at different quality levels. You can now see the Quality settings in the Inspector, and when you select a level it becomes highlighted. This is the quality you will see when you play the game in Unity.

It can be worth experimenting with this setting – especially if your computer is getting older. After all, you don't necessarily need your graphics using up all your processing power while you're working on your sound effects!

Assets Store

One advantage to using Unity as your Game Engine is that you have access to the Asset store. Here you'll find everything from basic sound effects, to 3D models and complete game templates. They range in price from free, up to hundreds of pounds/dollars.

However, be cautious when adding assets to a project that's still in progress, as there can be code clashes that break your game – especially if the assets haven't been updated to the same version of Unity you're using. Always back up your work before you play with any of these. With that warning in mind, they're definitely worth exploring and experimenting with.

And if you've created a library of your own sound effects, why not sell them on the Assets Store? If your background is in audio, you might have a wealth of experience in recording, editing, and creating sound – experience that many game makers do not have. People have a need for your skills, and there's a market for them here.

That's enough of an intro to Unity – you'll learn more by putting everything into practice. In Chapter 8, we'll start making our game world.

Level Building in Unity

Learning Outcomes

By the end of this chapter, you will be able to:

- Add a Terrain to the Unity Scene.
- Texture and shape the Terrain.
- Add trees and grass to the Terrain.
- Customize the sky in the Unity Scene.

The Terrain in Unity is a very efficient way to create landscapes in the Game Engine. To add one to your Scene, make sure you're in the Scene Tab, then go to GameObject > 3D Object > Terrain.

Move your mouse cursor over the Scene, and press [F] to center your view on the Terrain. Kind of boring so far – it's a white, feature-less square. However, as soon as the Terrain is selected, the Inspector gives you access to all the tools you need to shape and texture it.

Texturing the Terrain (Base Coat)

Let's start by adding some color, in the form of a Texture. The first texture you apply acts as a base coat, covering the entire Terrain (we'll paint over this later on).

Click on the Terrain tool that has an icon of a paintbrush. We can't do much here until we set the base texture. Click on the "Edit Textures" box, and choose "Add Texture."

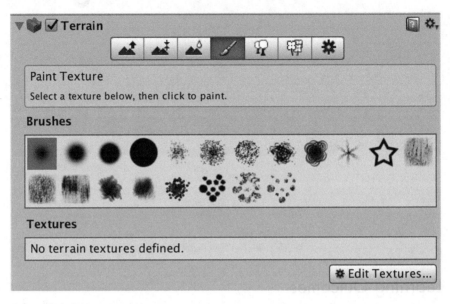

Figure 8.1 Paint texture

This opens up a floating window (you might have to look around for it, but it usually appears at the top left of the screen).

If your Terrain is set to the default setting, you will see two spaces to import Textures – "Smoothness" and "Normal" (if this appears any different, you'll need to change the Terrain Material type. This can be done from the Terrain Settings – look for the icon of a cog, and check that it's set to "Built In Standard").

The "Smoothness" box will set the basic reflective color (or "Albedo") of the Terrain.

"Normal" allows you to add a Normal (or "bump") map to the Texture. These are used to add visual embellishments such as small bumps and scratches, without actually changing the geometry of a model. This means models can have complex details, without needing to calculate and draw additional polygons.

For now, we'll just add a "Smoothness," so click on "Select" in the box. This shows you a list of all the 2D textures that have been added to the Project. (Most of these came with the "Environment" Asset package.) Choose one of these by double-clicking on it (I'm going to go for "GrassHillAlbedo"), and click on "Apply."

You will now see your Terrain has a texture applied to it. If you're zoomed out enough, you'll be able to see that it's repeated in a grid, but this won't be as obvious when we're playing the game (and we'll also be adding more textures on top).

The Textures box in the Inspector now shows your base texture (with a blue line underneath, indicating that this is currently selected). Double-click on this to re-open the Edit Terrain Texture box.

It's now possible to demonstrate the Metallic and Smoothness sliders (the texture needs to be rendered in the game to see the effect of these).

As the name suggests, as you turn up the Metallic slider, the texture appears more metallic – and the more light it will reflect back.

Smoothness determines the way that light is reflected. When this is turned down, the reflections are more diffuse (scattered). Turn this all the way up, and they will be more specular ("ray-like").

I'm going to leave these both at zero (but I can always come back and change them later on).

71

The Size parameters determine the size of the repeating grid, and you can also offset the patterns – useful when painting multiple textures on top of each other. Again I'll leave these set at their defaults.

Once we have a basic Texture applied, it will become far easier to see what we're doing when we start to shape our Terrain.

Resizing the Terrain

The Transform component Scale (and Rotation) values have no effect on the Terrain – they have to be changed from the Terrain Settings, so click on the tool with a cog icon to access these.

Towards the bottom, you'll find the Resolution, where you can set the width, length, and the maximum height.

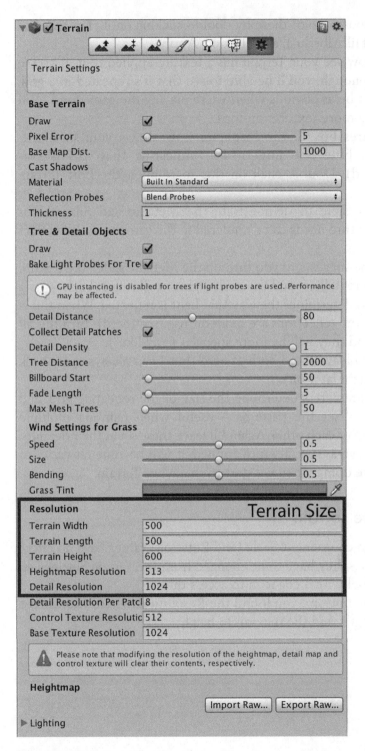

Figure 8.2 Terrain settings

If you're creating a game as a portfolio piece, you will want to cram everything close together (so any potential employer can see what you're capable of achieving in as short a time as possible). In this case, it will make sense to set the Terrain quite small.

For my game, I'm going to leave it set to 500 × 500, and I'm going to lower the height to 200.

Shaping the Terrain

The three tools on the left are used to sculpt and shape the Terrain. Start by selecting the middle one – Paint Height.

Paint Height (Flatten)

Under the Paint Height Settings, you'll see a parameter slider labeled "Height." Set this to 30, and click on "Flatten." This flattens the entire terrain height to 30 meters.

Now go to the top of the Inspector, and change the Y transform position to -30. This lowers the position of the Terrain by 30 meters.

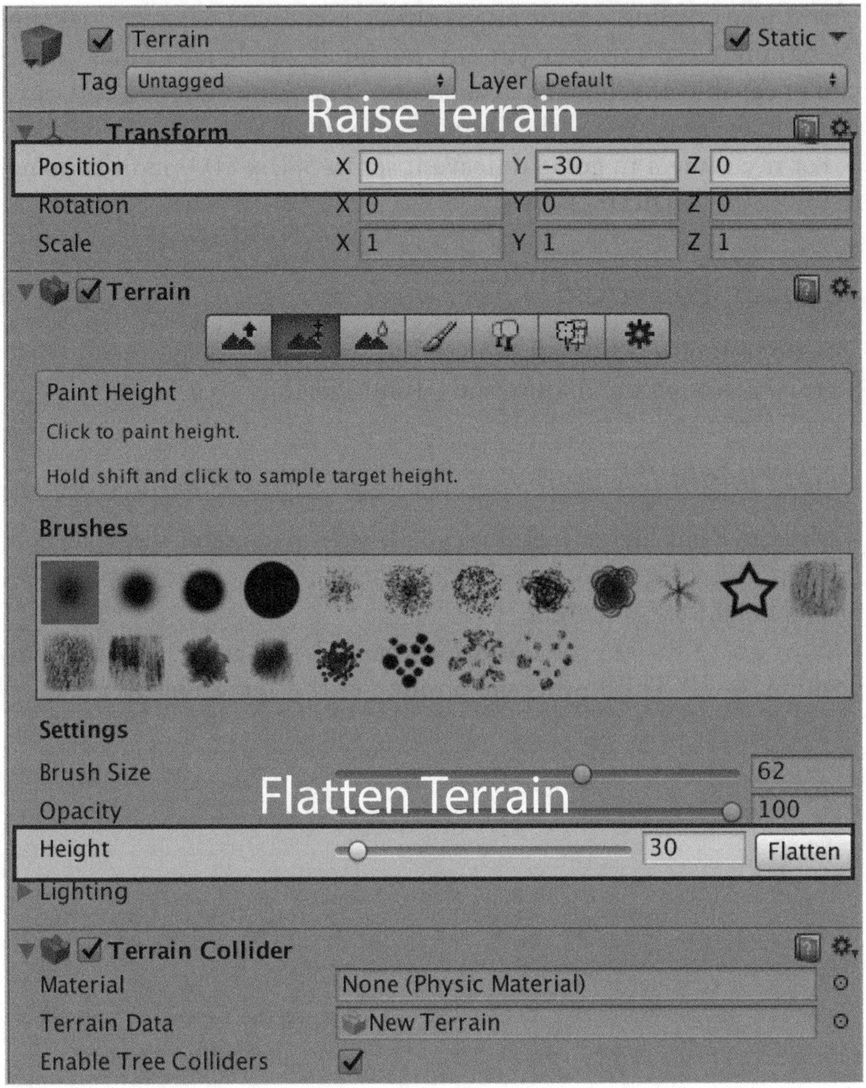

Figure 8.3 Raise Terrain and Flatten Terrain

If you look at the Terrain in the Inspector, these two actions appear to have cancelled each other out. So what was the point?

Well, it is not possible to have a negative Terrain height. By starting off with a flattened offset, you can now dig down into the Terrain by up to 30 meters, and our base terrain is still at 0 in the Y axis.

If you ever need to dig any deeper, you can always use a larger offset value (however, when you flatten a Terrain, you will lose any existing sculpting and shaping).

We'll come back to the Paint Height tool in just a moment, but for now, change to the Raise/Lower Terrain tool.

Raise/Lower

With this tool selected, choose your brush shape and size (I'd recommend beginning with a brush without a sharp edge, set to about 75).

Opacity is the "intensity" of the brush. Too low, and it will take ages to make any changes to the Terrain. Too high, and you'll be creating mountains with a single mouse click. 64 is a good compromise to start with.

Click and drag on the Terrain, and you'll raise it up. Hold shift and click, and you're lowering it. Before you know it, you're creating your first landscape.

Experiment with different brush shapes, sizes, and settings – it's easy to spend hours on this. I like to think of it as Minecraft for grown-ups… Don't forget that your player will need to navigate this landscape, so go easy on the mountains!

Paint Height (Sampling Terrain Height)

Switch back to the Terrain Height tool. This works in a similar manner to the Raise/Lower Terrain tool, but brings it to a defined height.

Hold shift, and click on a section of the Terrain. You have now sampled the height value (and this should be updated in the tool Height parameter). Now clicking on the Terrain will raise or lower it to this height.

This is very useful for creating flat areas (e.g. to place our models on), and pathways. Rather than leaving a player lost, a well-designed game will flow from one area to the next. Subtly shaped and textured pathways are a great way to do this, without the player feeling they're being led by the nose.

Smooth Height

This averages out the Terrain height, blurring any sharp edges.

Texturing the Terrain (Layering Textures)

Re-select the Paint Texture tool, click on Edit Textures, and add another Texture (I'm going to go for "CliffAlbedoSpecular").

Select the new texture so you can see a blue line underneath it. It's now possible to paint this texture on top of the "base coat" we added earlier.

Again, you can play with brush size and shape, but you also have two new parameters: Opacity and Target Strength.

Target strength is the maximum coverage that your texture will have. Set it to 100, and you will completely cover the existing textures.

Note that the texture layers remain separate, and can be changed at any time. If you paint over one layer with the Target Strength set to 100, then paint over again with a lower value, you will re-reveal the underlying layer.

Opacity is the intensity of your paint strokes. The lower it is set, the more strokes it will take to reach the Target Strength.

There are just two more Terrain tools to deal with – Paint Trees and Paint Details

Paint Trees

There's no prize for guessing what this allows us to do…

We'll need to start by adding tree types, so click on "Edit Trees," and choose "Add Tree."

This reveals the Add Tree menu. Just next to the Tree Prefab box is a circle with a dot in the center. Click here to reveal all the available trees…

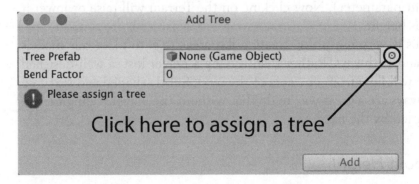

Figure 8.4 New tree

You'll see a few options here (which will give you a clue that you're not actually just restricted to trees), but I'm going to go for Palm_ Desktop, so double-click here, then click on Apply.

Take a look at the available settings. Turning up the Brush Size and Tree Density allows you to place entire forests in one go – though don't forget that your computer still has to render these! You can also randomize their height and rotation, to avoid planting blatant duplicates.

Paint Details

The most obvious use of this tool is when adding grass. Click on Edit Details, select "Add Grass Texture," then click on the circled dot to add a Detail Texture. Select one of the 2D assets (GrassFrond02AlbedoAlpha will do). Now change the Healthy and Dry colors, as the default can be quite garish, before clicking on Apply.

You can now paint patches of grass onto the Terrain (hold [shift] and click to mow the lawn). I'd proceed with caution – rendering grass can quickly eat up your processing power, and if your computer isn't high spec, I'd avoid grass altogether.

You can also use Paint Details to add decorations such as rocks and stones to the Terrain, but that's beyond the scope of what we'll be covering in this book.

The Sky

There are several options available to you here, and we'll be taking a look at two of them.

- In the Assets folder, right-click in the Project Tab gray-space to create a new folder, and name it "MyMaterials."
- Double-click on the folder to open it.
- Right-click in the Project Tab gray-space, and choose Create > Material.
- Name this "Sky1," and in the Inspector, set its Shader to Skybox > Procedural.
- Now drag the Material onto the Scene.

You can then adjust the setting for the sky in the Inspector.

77

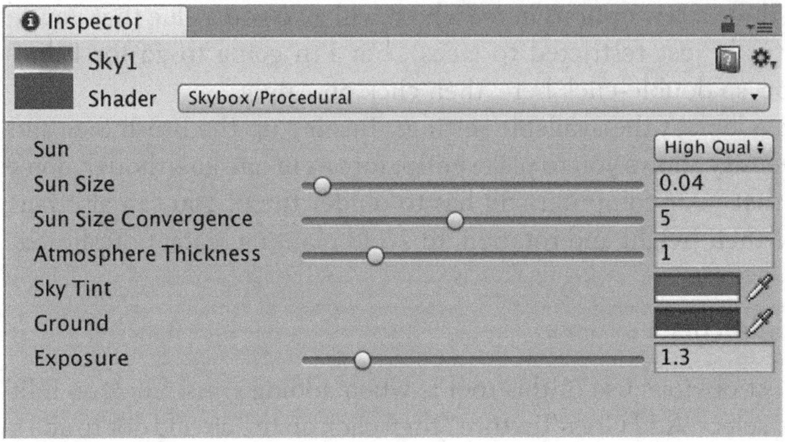

Figure 8.5 SkyBox – Procedural

However, this is somewhat limited – for example, there are no clouds or stars.

Another option is to use a Cubemap. However, you will need a Cubemap graphics asset. This is essentially an image of an unfolded cube, which is rendered around the Scene. You can find some of these free if you search online and in the Unity Asset store – and there's also a few of these included with the assets for this book.

- In the Assets folder, right-click in the gray-space to create a new folder, and name it "MyTextures."
- Double-click on the new folder to open it.
- Drag the Cubemaps folder from your computer Finder, into the MyTextures folder.
- Double-click on the Cubemaps folder to open it.
- Choose and select one of the Cubemaps.
- In the Inspector, set the Texture Shape to Cube and click on Apply.
- Navigate to the MyMaterials folder.
- Create another Material, name this "Sky2," and set its shader to Skymap > Cubemap.
- Drag the Sky2 Material into the Scene.
- Click on the "Select" option of the Cubemap (HDR) Inspector box of the Sky2 Material.
- Double-click to select your available Cube texture.

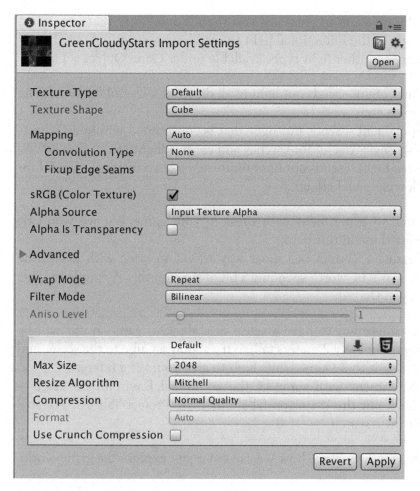

Figure 8.6 Sky – Set Cubemap

You now have a customized sky in your Scene.

Lighting

We already have a Directional Light in our Scene (what's sometimes known as an infinite light), and it represents our sun.

Its size and position in the Scene doesn't matter, but we can adjust its rotation to control the direction of shadows. As it's a parallel light source, all its shadows will be in the same direction – as opposed to a single point light, where the shadows will radiate away from the source.

If there's a sun in your sky, you should make sure that you adjust the angle of the Directional light to match its position.

There are other light types available under GameObject > Light:

- Point Light – sends out light in all directions. Useful for simulating lamps etc.
- Spot Light – sends out light that is constrained within an adjustable angle. Useful for torches and car headlights etc.
- Area Light – sends out light uniformly from a rectangle. Useful for glowing panel effects.

Feel free to experiment with these in your game, but we won't be covering these in this book.

Creating a Terrain is a great way to get to grips with navigating Unity, so it'll be worth spending a bit of time on this. A couple of hints to get you started:

- It's important to have a sense of scale. Our game avatar will be approximately 2 meters tall, so try adding a Cube to the scene, and scaling it to 2 Game Units high to give yourself a reference.
- Experiment with some of the additional Environment assets – check out the Water prefabs under the Project Tab "All Prefabs" Favorite search.

In Chapter 9, we'll see how we can enter and explore the game world.

Exploring the Game Environment

Learning Outcomes

By the end of this chapter, you will be able to:

- Use the WASD controls to play the game.
- Customize the First Person Character controls.
- Change the Unity preferences to add a color tint while the game is playing.

Hopefully you've had fun creating your Terrain. Now it's time to see how we can explore it. The Unity Standard Assets include a ready-to-go playable first-person character. In this chapter, we'll see how to add this to our game, and walk around the game world.

Take a look in the Project Tab "All Prefabs" Favorite search, and you'll see a blue cube named "FPSController."

Prefabs are GameObjects – you can think of them as presets (though, as we'll see later, instances of a Prefab in a game have linked properties). Drag the FPSController Prefab onto your scene – it should automatically rest on the Terrain.

Your game is now playable! Just press the Play button at the top of Unity (or use the [cmd]+[P] (Mac) / [ctrl]+[P] (Windows) shortcut). It uses the standard WASD controls:

- Use the mouse to look around.
- Use [W] and [S] to move forward and back.
- Use [A] and [D] to strafe left and right.
- Hold [shift] to run.
- Use the [spacebar] to jump.
- Press [esc] to regain control of the mouse cursor.
- Press the Play button (or use the shortcut) again to stop the game.

Let's take a closer look at what you've just added to the Scene. It contains a parent GameObject called "FPSController," and a child called "FirstPersonController." The FirstPersonCharacter contains:

- A Transform component (remember, there will be one of these on all GameObjects – see Chapter 7).
- A Camera component. This acts as the "eyes" for our game.
- A Flare Layer. These allow lens flare to be added (if you want to go all JJ Abrams on your game).
- An Audio Listener. This acts as the "ears" for Unity's built-in Audio Engine. We will be replacing this with an FMOD Listener.

This means our game currently has two cameras – the Main Camera that comes by default when a Scene is created, and the FirstPersonCharacter camera that you just added. While there are often reasons to use multiple cameras, we won't be needing both in our game. Right-click on the Main Camera in the Hierarchy and delete it.

The FPSController GameObject is more complex, and consists of:

- A Transform Component (see Chapter 7).
- A Character Controller. This allows you to control how the character will interact with the game environment and to set their eye height.
- A Script called "First Person Controller." This enables us to control and interact with the character (we'll be looking at this in detail in Chapter 10).
- A Rigidbody. This allows the GameObject to interact with the Unity Physics Engine (see Chapter 11).
- An Audio Source. This would allow the component to emit noises via the Unity Audio Engine.

If changes are made to GameObjects while the game is playing, as soon as it stops, they are reset.

- Press Play to start the game.
- Walk, jump, and run around the Scene.
- Pause the game (you can press [esc] to regain your cursor, then press the Pause button, or use the [cmd]+[shift]+[P] (Mac) / [ctrl]+[shift]+[P] (Windows) shortcut).
- Make changes to the First Person Controller component settings (for example, changing the Jump Speed to 80 will give your avatar superhuman powers).
- Un-pause the game (press the Pause button again, or use the [cmd]+[shift]+[P] (Mac) / [ctrl]+[shift]+[P] (Windows) shortcut).
- Experiment with the new settings.
- Press Play to stop the game. Your First Person Controller component parameters will reset to their original values.

This provides a safe way to try out changes to your game, without the risk of breaking it. However, it can be an infuriating way to lose work if you forget the game is paused.

To help avoid making this mistake, we can customize Unity's settings. Go to Unity > Preferences, and select the Colors Tab. You can then change the Playmode tint. This adds a color tint to the Unity interface while the game is playing, hopefully making it more obvious. I tend to go for a subtle pale green – though a deep red may be far more blatant, it can be quite distracting and tiring on the eyes!

Now you can test your Terrain, it's probably worth going back to ensure that your player can move around. You can also customize your FPSCharacter settings to suit the environment – for example, the Character Controller Slope Limit determines the steepest angle that the FPSCharacter can walk up.

In Chapter 10, we'll be taking our first look at code, and integrating our FMOD and Unity projects together.

83

10 Integrating FMOD and Unity

Learning Outcomes

By the end of this chapter, you will be able to:

- Integrate FMOD and Unity.
- Modify the FPSCharacter script to make it trigger FMOD Events.

If you've been playing your game, you'll have noticed that Unity is already making Footsteps, Jump, and Land sounds. These are part of the FPSCharacter prefab, and make use of the built-in Unity Audio Engine. In this chapter, we're going to replace these with the FMOD Events we created in Chapter 4.

When you downloaded FMOD, you were asked to save the UnityPackage to a safe place. This package contains everything we need to get the two applications working together.

Go to Assets > Import Package > Custom Package. Navigate to where you saved the UnityPackage, and click on "Open."

We'll need all of the package contents, so leave everything selected, and click on Import at the bottom right.

At the very bottom of Unity, you'll see a red error message, telling you that the FMOD Studio Project path has not been set (you can also see a slightly more detailed version of this message in the Console Tab).

Don't worry – this was entirely expected. We have not yet told Unity which FMOD project that we want it to use.

If you look at the top bar of Unity, you can see that we now have a new menu: FMOD.

Go to FMOD > Edit Settings, and the Inspector now gives us the option to set the Studio Project Path.

There are three options that need to be set according to the resources provided by the audio designer to the game developer.

You have the complete FMOD project, so leave this set to Project, and click on "Browse."

Navigate into your FMOD project folder, and select the .fspro file (mine is called FMOD_MyFPSGame.fspro).

That's all it takes. Your FMOD and Unity projects are now integrated.

At the moment, Unity is configured to use its own Audio Engine as well as FMOD. To prevent this, go to Edit > Project Settings > Audio, and in the Inspector, tick the "Disable Unity Audio" box.

Next, we need to place our listener in the game. These are FMOD's "ears," and it makes sense to place these on the same GameObject as the camera. In the Hierarchy, select the FirstPersonCharacter (remember, this is a child of the FPSController).

In the Inspector, you can see that there's already an Audio Listener component, but this belongs to Unity's built-in Audio Engine.

Click on "Add Component" and select FMOD Studio > FMOD Studio Listener (note that the order of Components is not alphabetical, so you'll have to hunt for this).

The existing Footsteps, Jump, and Land sounds are triggered by the FirstPersonController script. To replace these with our FMOD Events, we're going to need to modify the code.

You can either find the script in the Project Browser (the "All Scripts" favorite search will be useful), or you can select the FPSController, and then find it in the Inspector.

Double-click on the script to open it up in your IDE.

If you're working on a Mac, you may see a warning that your code endings need to be changed – just press OK to allow Visual Studio to fix them for you (don't worry too much about what this is doing – it's related to the different ways that Windows PCs and Macs represents line endings/new lines).

If you've not had any previous experience with coding, then you're probably a little apprehensive about the next few steps. If you want to just focus on the audio design side of things, you can simply follow the step-by-step guide. For those of you who require more background information, the Walkthrough boxouts go through exactly what the code is doing.

Code Terminology

Variables

Variables are how we represent our data values. Before we use a variable, we need to "declare" it. This means that we give it a name, and state its type.

It's common practice to declare all of your variables all together towards the start of the script (rather than place them where you need them in the code). You can see this on lines 13 to 43 in the FirstPersonController script.

Let's take a closer look at line 33:

```
33          private float m_YRotation;
```

Figure 10.1 FPS code line 33

- private: this means the variable cannot be accessed from outside the code. We'll be setting some of our variables to public. This will allow us to change values from the Unity Inspector, rather than having to open the IDE.
- float: this determines the that the data type is a floating point number – see below.
- m_YRotation: this is the name of the variable. Variable names should always start with a lower-case letter.

Types

Types are data formats. Some available types include:

- int: int values are whole numbers (integers). 3 is an example of an int value.
- float: float values are numbers, and they can be non-integers (i.e. they can have decimal points). 3.14f is an example of a float (floating point) value (the f is necessary to distinguish between another type called "double").
- double: similar to float, but can represent longer number values. 3.14 is an example of a double value.
- bool: can represent true/false values for logic operations (short for boolean).
- char: can represent a single character (number, letter, or symbol). '!' is an example of a char value.

- string: can represent a series of letters, numbers and characters (e.g. words and sentences). Written inside quotation marks, e.g. "this is an example of a string value".

Functions

A function is a section of script. It can be executed again and again. If you write code that performs a series of operations, you don't need to copy and paste the entire code if you need to perform the same operations again. You can simply call (or "execute") the function.

Functions in C# start with an upper-case letter, and are followed by brackets – for example, PlayFootStepAudio().

Take a look at lines 163 to 177 in the FirstPersonController script.

```
163    private void PlayFootStepAudio()
164    {
165        if (!m_CharacterController.isGrounded)
166        {
167            return;
168        }
169        // pick & play a random footstep sound from the array,
170        // excluding sound at index 0
171        int n = Random.Range(1, m_FootstepSounds.Length);
172        m_AudioSource.clip = m_FootstepSounds[n];
173        m_AudioSource.PlayOneShot(m_AudioSource.clip);
174        // move picked sound to index 0 so it's not picked next time
175        m_FootstepSounds[n] = m_FootstepSounds[0];
176        m_FootstepSounds[0] = m_AudioSource.clip;
177    }
```

Figure 10.2 FPS code line 163–177

This is where the function is declared – i.e. where we state what it does. Line 163 tells us:

- private: this function cannot be accessed from outside the code (as opposed to public).
- void: we do not expect any data output from this code (as opposed to defining the type of data output, e.g. integer [int]).
- PlayFootStepAudio(): this is the name of the function.

Immediately after this comes an open curly bracket ({).

If you select this, your IDE will highlight the close curly bracket (}) on line 177. Curly brackets separate different sections of code, so everything between these is part of the PlayFootStepAudio function.

This code checks to see that the player is on the ground, and if they are, it chooses and plays a random footstep sample.

If you look at line 159, this is where the PlayFootStepAudio function is called. This line is inside another function, which determines how often footstep sounds are played.

```
159          PlayFootStepAudio();
```

Figure 10.3 FPS code line 159

It would be possible to copy and paste lines 165 to 176 over line 159, and the game would still work (in fact, as this is the only place that the PlayFootStepAudio function is called, this would not actually make any difference).

Class

A Class is a collection of code (including functions) that acts like a blueprint. Once a Class has been written, it can be used multiple times – you do not have to re-write the code every time that it's needed. When a function is part of a Class, it is called a Method.

At line 11, you can see that this script derives from a class called "MonoBehaviour." This is the base class for all Unity scripts, and gives us access to some essential coding tools.

For example, lines 46 to 58 pertain to a Method called "Start()." Everything between its curly brackets is executed when the game starts, and is used to set the initial state.

As Start() is part of the MonoBehaviour class, the script did not have to directly include the code that defined the trigger condition of "when the game starts."

At the top of the page, you can see several items listed as "using" (for example, "using UnityEngine;"). This means that this script can also use their Methods.

```
1 using System;
2 using UnityEngine;
3 using UnityStandardAssets.CrossPlatformInput;
4 using UnityStandardAssets.Utility;
5 using Random = UnityEngine.Random;
```

Figure 10.4 FPS code line using

(Note: Strictly speaking, in Unity, all scripts are classes, which means that all functions are actually methods… This can cause some confusion, so we'll be sticking to the above terminology.)

Scroll down to line 87, and you'll see the PlayLandingSound function. We're going to replace the contents of this function with a single line of code, which will trigger the FMOD Land Event.

Select everything between the curly brackets (lines 89 to 91), and replace them with the line:

FMODUnity.RuntimeManager.PlayOneShot ("event:/Player/Vocals/Land");

Before

```
87          private void PlayLandingSound()
88          {
89              m_AudioSource.clip = m_LandSound;
90              m_AudioSource.Play();
91              m_NextStep = m_StepCycle + .5f;
92          }
```

After

```
87          private void PlayLandingSound()
88          {
89              FMODUnity.RuntimeManager.PlayOneShot ("event:/Player/Vocals/Land");
90          }
```

Figure 10.5 FPS code line 89–91 – before and after

If you're copying and pasting code, be very careful with quotation marks. Curly quotation marks are not treated the same as straight ones. If this is done correctly, Visual Studio will show the quotation marks and their contents in orange (this color may be changed in the application preferences).

The downloadable assets for this book include a text document called FMOD_FirstPersonControllerCodeForUnity. This is a plain text document (i.e. all formatting has been removed), so you won't have this issue.

However, it's essential to make sure you include the semicolon at the end!

Statements

A Statement expresses the action that code performs. Statements in C# can be single-line or blocks.

- A single-line statement in C# must end in a semicolon.
- A statement block is enclosed between curly brackets.

Leaving out semicolons and curly brackets are easy mistakes to make, and should be the first thing to check if your code isn't working the way you expect it to.

Now scroll down to line 135–139 to find the PlayJumpSound function. Paste over its contents with:

FMODUnity.RuntimeManager.PlayOneShot ("event:/Player/Vocals/Jump");

Before

```
135    private void PlayJumpSound()
136    {
137        m_AudioSource.clip = m_JumpSound;
138        m_AudioSource.Play();
139    }
```

After

```
135    private void PlayJumpSound()
136    {
137        FMODUnity.RuntimeManager.PlayOneShot ("event:/Player/Vocals/Jump");
138    }
```

Figure 10.6 FPS code line 135–139 – before and after

All that remains is to modify the PlayFootStepAudio function, which you'll now find at line 160.

We need to keep some of this: lines 162 to 165 prevent the Event from triggering when the player is in mid-air. Highlight lines 166 to 173, and paste over the code:

FMODUnity.RuntimeManager.PlayOneShot ("event:/Player/Footsteps/FootstepsGround");

Before

```
160    private void PlayFootStepAudio()
161    {
162        if (!m_CharacterController.isGrounded)
163        {
164            return;
165        }
166        // pick & play a random footstep sound from the array,
167        // excluding sound at index 0
168        int n = Random.Range(1, m_FootstepSounds.Length);
169        m_AudioSource.clip = m_FootstepSounds[n];
170        m_AudioSource.PlayOneShot(m_AudioSource.clip);
171        // move picked sound to index 0 so it's not picked next time
172        m_FootstepSounds[n] = m_FootstepSounds[0];
173        m_FootstepSounds[0] = m_AudioSource.clip;
174    }
```

After

```
160    private void PlayFootStepAudio()
161    {
162        if (!m_CharacterController.isGrounded)
163        {
164            return;
165        }
166        FMODUnity.RuntimeManager.PlayOneShot ("event:/Player/Footsteps/FootstepsGround");
167    }
```

Figure 10.7 FPS code line 166–173 – before and after

The script has now been modified to work with FMOD – and we could leave it like this, and it will work perfectly well. However, you will continue to get amber Console alerts telling you that "The private field 'UnityStandardAssets.Characters.FirstPerson. FirstPersonController.m_AudioSource' is assigned but its value is never used." This is simply telling you that there is a defined parameter in the script that is never used.

You should find this at line 43 (private AudioSource m_AudioSource;) and line 56 (line 55 once you remove line 43: m_Audi oSource = GetComponent<AudioSource>();). Delete these lines, and you will no longer get the warning.

(You are likely to still see other amber alerts, letting you know that certain elements are obsolete. Don't worry about these – some of the default assets still contain Components that are being gradually phased out, and Unity is just letting you know that you should avoid using these.)

Save your script, return to Unity and play the game. Hopefully, you will now be able to hear your own Events triggering!

If you want to tidy things up, you can uncheck the boxes on the FirstPerson Character Audio Listener component and the FPS Controller Audio Source component to de-activate them.

You'll also be seeing the FMOD debugger window in the top corner of the game screen. This can be useful to see what's going on, but if you want to get it out of the way, go to FMOD > Edit Settings, and set Debug Overlay to Disabled.

You now have a very simple playable scene, complete with basic sound effects. However, it's not yet much of a game. In Chapter 11, we'll be adding a GameObject for you to interact with.

Interactive Game Objects

Learning Outcomes

By the end of this chapter, you will be able to:

- Add a GameObject to the scene that interacts with the Unity Physics Engine.
- Add a material and texture to a GameObject.

What game would be complete without boxes and crates? From breaking them open for wumpa fruit in *Crash Bandicoot*, to taking cover behind them in *Half Life*, and hiding underneath them in *Metal Gear Solid*, they've long been standard video game props.

In this chapter, we'll use Crates to demonstrate adding interactive objects to the game world, and see how to apply textures and materials.

With the exception of the Terrain, the 3D modeling capabilities of Unity are very limited. If you wish to create your own models, you'll need to use dedicated software such as Maya or Blender.

However, it is possible to create some primitive shapes directly inside Unity. These are very basic objects that can be used for simple models and placeholders. These are:

- Cube
- Sphere
- Capsule
- Cylinder
- Plane
- Quad

We're going to use a Cube as the starting point for our Crates.

In Unity, go to GameObject > 3D Object > Cube. Name the Cube "Crate."

Use the Move tool to position the Crate somewhere close to the FPSController, about 1 Game Unit above the Terrain (this will make it easier for us to test).

Play the game, and try to interact with the Crate. You'll find that it's an immovable object, floating in space.

Stop the game, and take a look at the Crate Components in the Inspector.

As with all GameObjects, we have a Transform Component.

Next we have a Mesh. This is a collection of interconnected polygons (usually triangles), which defines the shape of the GameObject.

In the early days of game design, polygon count was a major consideration. The more polygons that are used, the more processing power is needed. This is no longer as critical (aspects such as lighting and shading are far more relevant), but it's still good practice to keep this as low as possible (without sacrificing too much detail). To give you a ballpark figure, most in-game characters have a polygon count of around 30,000 to 50,000 (and 100,000 polygons is by no means unheard of).

You can view your polygons in Unity by changing the Scene View setting. You'll find this at the top left off the Scene Tab – it will be set to "Shaded." Change this to Wireframe, and you'll see your Scene is made up of hundreds of triangles.

Figure 11.1 Wireframe

Mesh view can be useful for aligning GameObjects (and it significantly takes much less processing power to display), but for now, change back to Shaded.

Further down, you can also see a Mesh Renderer. This allows Unity to actually draw the GameObject. Try unchecking the tick box, and the Crate will still have a physical presence in the game, but you won't be able to see it. This can be useful if you want to add invisible barriers and walls to a scene.

There's also a Box Collider. Colliders define the mesh (shape) of GameObjects when calculating physics. As we have a Collider component attached, the Cube is solid, and will interact with other GameObjects. If you disable this component, then you'll be able to walk straight through the Crate.

Finally, we have a Material. This will be set to the default Material (a boring off-white/gray). Materials are used to add textures and set the shading properties of GameObjects.

You can see that the parameters are grayed out, so they can't be changed from here.

Let's make our Crate look more interesting.

95

Navigate to the MyMaterials folder you created in Chapter 8, and right-click in the Project Tab gray-space to create a new Material. Name this "CrateMaterial."

At the bottom of the Inspector, you can see a preview of the Material – and you can click and drag on this to view it from different angles. However, it will be better if we apply the Material to the Crate before we make any changes – this way, we can see how it will look under the scene lighting conditions.

Drag the CrateMaterial onto the Crate (it's safest to do this using the Hierarchy).

In the Inspector, you can see that this has replaced the Default Material, and you can now access the Material properties.

A quick warning though – if you make any changes here, it will affect all GameObjects that are using the same Material (this won't actually cause us any problems here, as we have only used the CrateMaterial once, but it's something you should definitely be aware of).

If you click on the white box just to the right of the Albedo parameter, you can set a single color for the Material. This can be useful, but will give everything a very cartoony look. Have a quick play with this, and once you're finished, click on the cog icon at the top right of the component to reset the Material.

It's far more common to use Texture Maps. These are 2D graphics that are wrapped around the 3D GameObject.

UVing and Texturing

UVing
UVing is the process of "skinning" a 3D model, creating a 2D shape that can then have graphics painted onto them. The term comes from the two axes of the resulting output file (U and V). This can be quite a difficult process – especially if you have a complex mesh.

Most 3D design applications can do this semi-automatically for you, but for the best results you'll need to fine-tune these by hand.

Texturing
Texturing is the process of painting the colors and details onto the UV.

It's difficult to picture the effect of painting a 2D texture map on a 3D model, but it's also difficult to work on a 3D model on a 2D computer screen. This can be done in most 3D design applications, but I'd recommend trying out Mari (by the Foundry), which is designed specifically for this purpose. It's quite easy to get to grips with the basics, and probably the easiest way for you to customize your texture assets (The Foundry offer free student versions of nearly all their software packages).

A Cube is the simplest shape for us to deal with, and for something as basic as our Crate, we can use a square image that will be repeated on each face.

Import the CrateTexture.psd and CrateNormal.psd files into the MyTextures folder.

Now drag the CrateTexture.psd texture to the CrateMaterial Albedo box.

Your Crate now has a simple texture applied to it. Play the game, and have a walk around it.

Although we've applied a texture, it's quite obvious that this is a 2D graphic painted onto a 3D object.

To make this look less flat, we'll need to add 3D details. One approach to this would be to use a more complex 3D model. However, this would add significantly more polygons, which would then of course take more processing power.

Instead, we can use a Normal Map. This "fakes" 3D details, by modeling the way that light reflects in different directions on a bumpy surface.

Drag the CrateNormal.png texture onto the CrateMaterial Normal Map box. You will see a warning that the texture is not marked as a normal map, but don't worry about this – just click on "Fix Now".

Check out the Crate in the game, and you can see that it looks much more realistic and 3D.

Figure 11.2 Textured Crate

There are several more Material parameters that we won't be covering in this book. If you want to learn more about these, your first step should be to check out the Unity documentation. Click on the icon of a book at the top right of any component, and it will take you straight to the appropriate page of the Unity website.

Our Crate now looks the part, but it's still an immovable object. To make a Game Object interact with the Physics Engine we need to add a Rigidbody.

Select the Crate, and click on "Add Component" at the bottom of the Inspector.

We could navigate straight to the Rigidbody (it's under "Physics"), but if you don't know where to find a component, click next to the magnifying glass icon, and start to type in the name. "ri" is enough here to find what we're looking for, so click on the Rigidbody to add it to the Crate.

Now play the game, and you'll see the Crate fall to the ground. Walk into it, and you can push it around the Terrain.

It's worth experimenting with the Mass setting (which sets the weight in kilograms). Too light, and the Crates will roll out of the way when you brush slightly against them. Too heavy, and they won't ever move. I've found that 2kg is a decent compromise.

Prefabs

We now have a Crate in our game. However, we're going to need several of these.

It's possible to duplicate GameObjects (try right-clicking on the Crate in the Hierarchy). However, if you ever need to make any alterations to the Crates (for example, we will be adding an FMOD Emitter in Chapter 12), you would then have to go through them all one by one, applying the same changes to each of the copies.

Instead, we're going to create a Prefab.

- Create a new folder called MyPrefabs in the Project Tab Assets folder, and double-click to open it.
- Drag the Crate from the Hierarchy onto the Project Tab. This has now used the Crate GameObject as the basis for a new Prefab.
- You can now drag the Crate Prefab into the Scene as many times as you like.

99

Positioning GameObjects

If the Toggle Tool Handle position is set to "Center," the Prefabs will sit on top of any existing GameObjects (including the Terrain), so it's quite easy to place your Crates where you want them – even to stack them on top of each other.

If you're not happy with the placement of any of your Crates, you can use the Move tool to re-position them. However, unless the ground is flat, it will be easy to accidentally place them under the surface of the Terrain (or leave them floating in the air).

Not to worry – holding down [shift]+[cmd] ([shift]+[ctrl] on Windows) and clicking on the square in the center of the selection enables Surface Snapping – GameObjects will remain seated on the Terrain when they are moved. Try this on the original Crate to place it on the ground (remember, we originally set it hovering 1 Game Unit off the floor).

The Crate in the MyPrefabs folder can be thought of as the "master" Crate. If you select this and make any changes, the changes will be applied to all instances of the Prefab in the scene.

(Note – the Position and Rotation Transform master settings do not have any direct effect on the Prefab instances.)

If you select one of the Crates in the Scene (or Hierarchy), it is possible to make individual changes to a Prefab without affecting the other instances.

A Prefab has three additional options at the top of the Inspector: Select, Revert, and Apply. These allow us to make changes that can affect all of the other instances.

- "Select" will select the master Prefab in the Project Tab.
- "Revert" will reset any individual changes you have made to the Prefab instance, so it is now the same as the master.
- "Apply" will apply the individual changes to the master, allowing you to push these changes to all other instances of the Prefab.

100

Once you've placed a few Crates in your scene, it can be worth tidying them up, by placing them inside an Empty Game Object.

Go to GameObject > Create Empty. Name this "Crates," and in the Hierarchy, drag all of the Crate prefabs on top, making them its children.

This allows you to clean up the Hierarchy – clicking on the Crates disclosure arrow will hide them from view. It's also now possible to remove them from the Scene at once – select the Parent in the Hierarchy, and uncheck the active tick box at the top left of the Inspector.

Don't worry too much about the positions of your Crates for now – you'll be moving them around to suit the game environment. For now, make sure you have a few of these placed somewhere close to the FPSController. In Chapter 12, we'll be adding impact sound effects.

Crate Sounds

Learning Outcomes

By the end of this chapter, you will be able to:

- Add an FMOD Studio Event emitter to a GameObject that will trigger an Event when the GameObject collides with a Rigidbody.
- Add pitch randomization within an Event.

Our game now has an object that we can push around the world. In this chapter, we'll be adding impact sound effects.

There are sound effects for the Crate impacts included in the book assets, but this would be an ideal opportunity for you to create your own.

Recording Equipment

Recording sound effects at home is fairly straightforward – all you'll need is an audio interface, a microphone, and a DAW.

Interface
As long as your audio interface has a mic input, it will probably do the job. However, some of the lower-budget interfaces have noisy

preamps. This results in "hissy" recordings – especially when you're trying to capture quiet sound effects. If you're looking at getting an interface, you won't go wrong looking at what's offered by Audient and Focusrite.

Microphone

When recording sound effects for film and TV, it's very common to use shotgun microphones (notably the Sennheiser MKH 416). However, the main reason for this is that these are the same models used on set/location. This ensures a tonal consistency, which helps the sound effects to blend with the original recordings.

We don't have this issue in our game, so we're free to use whatever we like. I have a variety of microphones (including dynamic mics, large and small diaphragm condenser mics, contact mics, lavaliere, and shotguns), and choose whatever is best for the job.

If you're looking for a good all-rounder, then I'd recommend starting with a large diaphragm condenser. This can be used on everything including vocals, instruments, foley, and other sound effects.

Digital Audio Workstation (DAW)

Any DAW will be suitable. Pro Tools is pretty much the film sound industry standard, but in game audio you'll often find people using Nuendo, Reaper, and Audacity. Stick to whatever you're comfortable with.

When recording in the studio, you should attempt to capture clean, reverb-free sounds. If your sound effects already have reverb, it can be difficult to blend these realistically into the game environment. The human ear can interpret a remarkable amount of information from reverb, and if the reverb doesn't match the graphics, it will affect the player's immersion.

This is why a dedicated sound effects studio is very dry (reverb-free), with the walls and flooring covered with porous materials.

In a home setup, acoustic foam panels can be placed around the sound source to dampen reflections – pillows and blankets will do in a pinch.

Recording outside and on location is more of a challenge. You can take a laptop and interface with you, but this is obviously going to be awkward. Location recorders are a much simpler option.

Location Equipment

Recorders

You can pay thousands of pounds for high-end location recorders, but it's possible to get very usable results without breaking the bank. Tascam and Zoom offer much more affordable solutions.

Some of these come with built-in microphones. While this is nice and convenient, they are almost always stereo mic arrays. Stereo recordings are great for ambiences and atmospheres, but sound effects should almost always be mono. It may be possible to switch off one of the mic capsules for mono recording, but I prefer to use a recorder with XLR mic inputs.

Microphone preamps on budget location recorders tend to be quite noisy, so I use a Triton Audio FETHead. It's essentially an amplifier that goes between the mic and the mic pre on the recorder. It makes a huge improvement in quality – but make sure you get the right model to go with your microphone.

Microphones

Wind and background noise are significant issues when recording on location.

The more directional a microphone's polar pattern, the more it will suffer from wind noise. This would mean that omnidirectional mics are ideal. However, these mics are (by definition) equally sensitive to sound originating from all directions, so will pick up significant background noise.

This means that we'll probably need to use a shotgun mic with a wind shield. If you're on a tight budget, you're looking at something along the lines of a Røde NTG2.

Miscellaneous

You'll need some way to hold the microphone – either a boom or hand-grip. Most location recorders can be mounted on a camera stand, which can be useful when recording long ambiences and room tones.

Don't forget headphones! You'll need a pair that provides enough isolation for monitoring, and are sturdy enough to survive the beating they'll take. I use Sennheiser HD25s, but I've also had success using a pair of in-ear headphones covered by a set of cheap ear defenders.

Audio restoration software can be invaluable to clean up and rescue recordings. Izotope RX has been a lifesaver on several occasions!

103

- Gather all of your Crate impact samples.
- Open up your FMOD project and import them in.
- Create a new folder and name it "Props."
- Open the folder, and create a new 3D Event.
- Name this "CrateImpacts."
- Select all of the Crate impacts in the Audio Bin, and drag them to the CrateImpacts Audio Track, to create a new Multi Instrument.

Play the Event a few times, and you should hear it shuffle between the different samples. It won't take long before you'll start to notice repetition.

We could take a similar approach to what we did for our footsteps, and layer multiple samples on top of each other. However, this would require even more sound assets and processing power. Instead, we're going to add pitch randomization.

Select the Instrument, and at the bottom left of the Deck you'll see the Pitch parameter. This works by changing the playback speed – slowing it down lowers the pitch.

Turn it all the way down to -24 semitones and play the Event. You'll notice that the end of the samples will be truncated. This is because the Instrument Trigger Region is not long enough to cover the entire duration of the sample (lowering a sample by 24 semitones makes it four times longer).

While we're unlikely to slow the Crate impacts by this much, it (hopefully) illustrates the problem. Fortunately, this has an easy fix – click on the Async button to give each sample its own independent timeline.

Double-click on the pot to reset the pitch.

Now right-click on the pot, and choose Add Modulation > Random.

This adds a new parameter in the Deck – pitch randomization.

This sets the maximum amount of randomization. Around seven semitones should work for the CrateImpact Event, but feel free to experiment with this for yourself.

If you now examine the Pitch pot, you'll notice a green bar around it. This shows the randomization pitch range. You can use this control to set the pitch center – for example, turning it anti-clockwise can ensure that the samples never play higher than their original pitch.

Pitch Shifting

Tonal sounds (such as vocals) have components called formants, which are constant, regardless of the pitch produced. When the playback speed is changed, it has the effect of altering the formants. This is what causes the "chipmunk" effect. This can be extremely noticeable, so we especially avoid pitch randomization on vocal samples.

This is not as obvious on percussive and atonal sounds (such as our Crate impacts), so we can get away with cranking it up quite high before the pitch-shift artifacts become too obvious.

However, if you are editing samples that you know will be lowered in pitch, you should take care to trim the start of the sample as short as possible, as any delay will be exaggerated, giving the appearance that the Event is triggering late.

Because pitch shifting is commonly used in game audio, it is common practice to record audio samples at higher sample rates (96kHz and higher). This allows frequencies higher than 20kHz to be captured. These will then be pitch shifted down into the audible spectrum when the recording is played back at a slower speed.

105

Our Event is pretty much ready to test out in the game, so right-click in the Events Tab to assign it to a Bank. Let's create a dedicated bank for this Scene, so right-click on the Event, choose Assign To Bank > New Bank, and name the Bank "BattlegroundBank." Save and Build the project, and return to Unity.

Find and select one of the Crates. In the Inspector, click on "Add Component," and navigate to FMOD Studio > FMOD Studio Event Emitter (or start typing FMOD in the search box).

In the Event Emitter Component, set the Play Event to "Collision Enter" (you might have to click on the disclosure arrow to view the parameters). This will cause the Event to be triggered whenever the Crate collides with another object (provided at least one of the GameObjects has a Rigidbody attached) – for example, other Crates and the Terrain.

Now click on the magnifying glass, then find and select your CrateImpact Event.

So far, this will have only added the FMOD Studio Event Emitter to the selected Crate. At the top of the Inspector, click on "Apply" to make the changes to all the Prefab instances.

Play the game. Try running into the Crates and making them bounce around (it might help to play with the Rigidbody Mass settings). You should be able to hear the Crate impacts.

(Note: if you have touching Crates, you might find that they continuously trigger the CrateImpact Event – if you look closely, you'll see that they are also vibrating slightly. This is a slight bug. You can usually fix this by removing, then re-applying, the Rigidbody Components.)

You'll probably find that the impact sounds are too loud, and you'll want to experiment with the Distance Attenuation settings (don't forget, to apply your FMOD changes to the game, you'll need to Save and Build the FMOD project).

When you select a GameObject with a 3D FMOD Studio Event Emitter component, you'll be able to see two spheres surrounding it in the Scene. These illustrate the Min and Max Distance Attenuation (if you want to see how these change, you can check the "Override Attenuation" setting in the Event Emitter component, and alter the settings from inside Unity).

Play around with the CrateImpact Event in FMOD. Try additional layers, as well as trying out different pitch and pitch randomization settings, until you get to something you're happy with.

In Chapter 13, we'll be adding a rocket launcher, so you'll be able to shoot at the Crates for target practice...

Putting the "S" into FPS

Learning Outcomes

By the end of this chapter, you will be able to:

*   Add a firing rocket launcher to the game.
*   Add Rockets to the game that explode on contact with other GameObjects.
*   Add sound effects for the weapon firing, Rockets, and explosions.

As our game is going to be a First Person Shooter, we're going to need weapons. We'll keep things simple, and stick to a rocket launcher.

There's a file called RocketLauncher.fbx included in the book assets (you can use this, or see what you can find online or in the Unity Asset Store).

Create a folder in your Unity project called "MyModels," and drag the RocketLauncher.fbx file into Unity to import it into here (the fbx [filebox] format was originally created for motion capture, and is commonly used for all 3D data – including models. The textures and materials are included in the file).

Select the RocketLauncher.fbx file in the Project Tab, and look at the bottom of the Inspector to preview the model. You can see that it's currently gray, without any textures. Textures and materials are included in fbx files, but Unity won't automatically extract them.

At the top of the Inspector are several buttons. Click on "Materials." You're now given the option of Extract Textures, so click here. Click on the menu button to create a new folder for your textures, name the folder "RocketLauncherTextures," and select "Choose."

If you return to the MyModels folder and click on the disclosure arrow of the RocketLauncher, you can see that it's made up of several Components, including a Material. Select the Material, and in the Inspector you'll see that all of the parameters are grayed out. If you're happy with the way that things look, then this is fine, but if you want to change anything, we'll have to take one more step.

Re-select the RocketLauncher, go back to the Materials menu and click on "Extract Materials" (if you want to create another folder for this, then go ahead). Find and select the extracted Material (it will no longer be a component of the RocketLauncher model – it will be found wherever you just extracted it to – you will find the "AllMaterials" Favorites menu useful here); you will now have access to all of the Material settings in the Inspector.

Next, we'll place the rocket launcher in the game. It will need to be a child of the FirstPersonCharacter (i.e. a "grandchild" of the FPSController), so check that it is visible in the Hierarchy, and drag it on top.

We have to fine-tune the position of the RocketLauncher (at the moment it is slightly behind the camera). It's going to be tricky to do this in our current Unity view, so go to the top right, and change the Layout to "2 by 3," which shows the Scene and Game views simultaneously. However, it will also reset the Scene view position, so:

- Find the Hierarchy Tab (it's moved to the right of the Scene and Game Tabs).
- Click on the disclosure arrow to open the Scene.
- Select the FPSController.
- Mouse over the Scene window.
- Press [F] to focus on the FPSCharacter.
- Navigate thought the Hierarchy disclosure arrows, and select the RocketLauncher.

You can now play with the position and scale the RocketLauncher until it looks right. Once you're happy with this, change back to the Default Layout.

Try playing the game, and you'll find that the RocketLauncher is intangible – it goes through other GameObjects. This is because it does not yet have a collider.

The RocketLauncher is made up of two component parts – the Barrel and the Sight. Select the Barrel in the Hierarchy, and click on "Add Component."

If you type "collider" into the search box, you'll see that there are a number of available options.

If you choose Mesh collider, then the collider will take the same shape as the 3D model. This is great if you need to be extremely accurate with your collision detection, but takes more processing power. You could reduce this by using a simplified mesh that omits some of the finer details of the 3D model, but it's usually a better option to use primitive colliders – the capsule collider will do a reasonable job here.

When you add a capsule collider, you'll see that it automatically sizes to cover the entire model. While that's helpful, it's not ideal in this case – we only want it to cover the barrel of the RocketLauncher.

That's simple enough to fix – change the radius and center in the Inspector. One useful hint here – it's possible to click and drag on the parameter name to change the value. Try this out on the Radius setting.

That's probably good enough for now, but if you want to improve the collision accuracy, add more primitive colliders to cover the rest of the model.

That's the RocketLauncher done with – now we just need to create a Rocket. Again, we'll keep things simple, by using a primitive shape. Go to GameObject > 3D Object > Sphere. Name this "Rocket."

This is obviously too large, so play with the Scale until it looks as if it could come out of the barrel of the launcher. Switching to Isometric view (with the Scene Gizmo) and moving the Rocket around will help you to get a better sense of their relative sizes.

We will want our Rockets to interact with the game physics, so click on Add Component and add a Rigidbody. We don't want our Rockets to affected by gravity, so uncheck this option (though if you want your game to have weapon ballistics, leave this setting on).

109

There will be multiple Rockets instantiated in our game, so we'll be using Prefabs. Drag the Rocket from the Hierarchy into the MyPrefabs folder in the Project Tab.

We won't want the original Rocket in the Scene, so right-click on it in the Hierarchy and choose "Delete."

Next, we're going to set the position where the Rockets will fire from – this will obviously be at the mouth of the launcher.

Right-click in the gray-space of the Hierarchy and choose Create Empty. Name this "ShootFromHere," and move it just in front of the RocketLauncher (this can be quite fiddly, so you'll need to make use of the Scene Gizmo to get it right).

Now drag the ShootFromHere Game Object onto the Rocket Launcher in the hierarchy, so that it is a child of the RocketLauncher (i.e. a great-grandchild of the FPSController).

(Note – it would have been possible to make the ShootFromHere GameObject a child of the RocketLauncher when it was created. However, this would set its rotation the same as the parent – and if there have been any changes made here, this would result in the Rockets firing off in an odd direction. This can happen quite easily, as different 3D applications use different labels for their axis, and the models might have been rotated in Unity to compensate for this.)

That's got everything set in the Scene, so it's time to get coding. Create a folder called "MyScripts" in the Project Tab. Double-click on this to open it, then right-click > Create > C# Script.

Name this "ShootScript."

Naming and Renaming Scripts

When you create a new script, it immediately gives you the option to type in a name. This actually has an effect on the script file that is created – when you open the script in an IDE, you'll see a line of code towards the top that reads "public class NameYouGaveTheScript: Mo noBehaviour {"

The file and class names must always match (as mentioned earlier, Unity scripts are technically Classes), so if you ever need to change the name of a script, you'll also need to make the same change to this line of code.

Double-click on the script file to open it in the IDE.
Now write the following code:

```
1 using System.Collections;
2 using System.Collections.Generic;
3 using UnityEngine;
4
5 public class ShootScript : MonoBehaviour
6 {
7     public Rigidbody rocket;
8     public float speed = 40;
9
10    // Use this for initialization
11    void Start()
12    {
13        Cursor.lockState = CursorLockMode.Locked;
14    }
15    void FireRocket()
16    {
17        Rigidbody rocketClone = (Rigidbody)Instantiate(rocket, transform.position, transform.rotation);
18        rocketClone.velocity = transform.forward * speed;
19    }
20
21    // Update is called once per frame
22    void Update()
23    {
24
25        if (Input.GetButtonDown("Fire1"))
26        {
27            FireRocket();
28        }
29    }
30 }
31
32
```

Figure 13.1 ShootScript 1

(The code is also available in the ShootScript1 text document in the book resources, so you can copy and paste it over to avoid any typo errors.)

ShootScript Code Walkthrough (Version 1)

At the start of the script, two public variables are declared – a Rigidbody named Rocket, and a float named speed. These will allow us to set which GameObject is used for Rockets, as well as their speed.

As these variables are public, we can change these settings from within Unity – they will be accessible directly from the Inspector Component.

The speed variable already has a value of 40. This will be the default setting unless you change it in Unity.

The Start Method always runs when the script starts (in this case here, when the game begins). This single statement locks the mouse cursor to your game window. This is useful, as otherwise if it isn't over the Scene when you press Play, you won't be able to shoot until you bring it back and click in the Game window.

Lines 15 to 19 define the FireRocket function. We don't expect an output value from this, so it has a type of "void."

When the function is called, it instantiates an instance of the Rocket called "rocketClone" at the position of the GameObject the script is attached to (this will be the "ShootFromHere" empty GameObject). The rocketClone will then be fired forwards (at the velocity determined by the "speed" variable).

Then we have the Update method on lines 22 to 29. This is available because our script is part of the MonoBehaviour class. Anything in here is run once per frame – in this case, it checks if the left mouse button ("Fire1") is pressed, and if it is, the FireRocket function is called.

Save the script, and return to Unity.

Drag the script onto the "ShootFromHere" GameObject in the Hierarchy. If you select this, you can now see the attached script component in the Inspector, and access the two public variables (rocket and speed).

The easiest way to set the rocket variable is to drag the prefab into the parameter box. However, as soon as you try to navigate to the MyPrefabs folder to find the Rocket, the Inspector changes its focus.

Not to worry. With ShootFromHere selected, click on the icon of a padlock at the top right of the Inspector. This locks the Inspector to the current object, so you can find your Rocket prefab, and drag it in. Don't forget to unlock the Inspector once you're finished.

Play the game, and try out shooting at the Crates. You'll be able to move them around, but there are a few issues we'll need to fix.

First, you might have noticed that occasionally the Rockets pass straight through objects. This is because they're moving too fast. The game only checks once per frame to see if the Rocket's sphere collider has intersected with anything – and at the current velocity, it sometimes goes all the way through before the next frame.

To fix this, find and select the Rocket prefab in the Project Tab. Now change the Collision Detection from Discrete to Continuous. This will detect any collision, regardless of frame rate or velocity. Continuous detection uses more processing power, so we only use this setting when absolutely necessary.

Next, when we fire Rockets, they remain in the game until we press stop (you can see them arriving in the Hierarchy every time you fire one). Over the course of a game this could add up to hundreds of Rockets, each of which continues to use up processing power.

This won't be an issue if we hit anything (we'll set them to destroy on impact), but if we miss or fire into space, then they'll be there forever. Let's fix this with another script, which destroys the Rockets after a set time.

Navigate to your MyScripts folder, and right-click to create a new C# script called "DestroyRocket".

Double-click on this to open it up in your IDE, and add the following code:

```
1 using System.Collections;
2 using System.Collections.Generic;
3 using UnityEngine;
4
5 public class DestroyRocket : MonoBehaviour
6 {
7     public float lifetime = 3f;
8
9     // Update is called once per frame
10     void Update()
11     {
12         Destroy(gameObject, lifetime);
13     }
14 }
```

Figure 13.2 DestroyRocket (Time Only)

(This code is available in the DestroyRocket1 text file in the book resources.)

Save your script, return to Unity, and place the DestroyRocket script on the Rocket prefab (you'll have to lock the Inspector to the Rocket prefab again).

DestroyRocket Script Walkthrough (Time Only)

Quite a simple one here.

We have a public float value called "lifetime," with a default setting of 3 seconds.

Then, in the Update (once per frame) method, we have a line of code that destroys the gameObject that the script is attached to, once the lifetime value has been reached.

"Destroy" is a function that's available to us because our script is using the UnityEngine class (see line 3 in the code).

Note that line 12 says "gameObject" with a lower case "g" (as opposed to GameObject). This means that it will only affect the specific instance of the Rocket that it is attached to – not all of the Rockets in the game.

Try out the game, and you'll see that the Rockets will disappear after a few seconds. If you think that's too quick, then you can change the lifetime in the Unity Inspector.

Now we want to make the Rockets explode when they collide with anything. We're going to need to import another of Unity's default assets – Particle Systems.

- Go to Assets > Import Package > Particle Systems.
- Leave everything selected in the pop-up floating window, and click on "Import."

Return to the DestroyRocket script, and modify it as shown below:

```
1 using System.Collections;
2 using System.Collections.Generic;
3 using UnityEngine;
4
5 public class DestroyRocket : MonoBehaviour
6 {
7     public float lifetime = 3f;
8     public GameObject explosion;
9
10    // Update is called once per frame
11    void Update()
12    {
13        Destroy(gameObject, lifetime);
14    }
15
16    void OnTriggerEnter(Collider other) //Code is called when a collider or rigidbody touches the trigger
17    {
18        GameObject boom = Instantiate(explosion, transform.position, Quaternion.identity) as GameObject;
19        Destroy(gameObject); //Destroy the rocket
20        Destroy(boom, 3); //Destroy the explosion after 3 seconds
21    }
22 }
```

Figure 13.3 DestroyRocket Script (On Trigger Enter)

(This code is available in the DestroyRocket2 text file in the book resources – don't miss out line 8 "public GameObject explosion;.")

115

- Save the script, and return to Unity.
- Select the Rocket prefab, and lock the Inspector.
- Type the word "explosion" into the Project Tab search box, and you'll see a couple of scripts, as well as two explosion prefab particle systems.
- Drag the Explosion prefab into the Explosion box of the Destroy Rocket component.
- Now set the Rocket prefab Sphere Collider to "Is Trigger".

DestroyRocket Script Walkthrough (OnTriggerEnter)

OnTriggerEnter is a Method included in MonoBehaviour, and allows colliders to trigger code.

Our Rocket prefab already has a collider – however, unless this is set to "Is Trigger," it can't be used by the OnTriggerEnter Method.

We could have added a second collider to the Rocket prefab, but we won't need the impact of the Rockets to have a physical interaction with other GameObjects – the Explosion particle system has a built-in physics force that will do the job for us. Instead, we set the existing sphere collider to "Is Trigger."

When the Method is called, Unity will instantiate an instance of the explosion called "boom," and destroy the Rocket (lines 18 and 19).

After a few seconds it then removes the boom from the game (line 20).

You should now have a functioning rocket launcher in the game! Try it out, and you'll note that all that's missing are the sound effects.

Save Unity, and open your FMOD project (the shortcut [cmd]+[shift]+[O] (Mac) / [ctrl]+[shift]+[O] (Windows) opens the previous project).

In the Events Tab, right-click to create a folder called "Weapons," and add another folder inside here called "Explosions."

Now add a new 3D Event called "ExplosionSFX." You'll need some explosion samples, so either find some of your own, or use the files included with the book assets. Import these to FMOD, and use them in a Multi Instrument.

We'll need some variation to the explosions, so experiment with pitch randomization and layering until you're happy with the way it's sounding.

One great way to add realism to your explosion sound effects is with the sound of rubble landing afterwards. Right-click on the existing Audio Track header to create another, and use the SFX_Explosion_Rubble files in a new Multi Instrument (it will help again to set this to Asynchronous).

If you preview the Event, you probably won't be able to hear the rubble – it will be masked by the explosions. We could move it later in the Timeline, but it wouldn't sound right if the rubble is always heard with the same delay after every explosion.

Select the Rubble Instrument, and click on the disclosure arrow to view the Trigger Behavior settings in the Deck. By default this is set to Tempo, so click on the button to change this to Time.

You now have a Delay Interval, where you can set the soonest and latest the rubble will play after the Event is triggered. Experiment with this – I've found that between 500ms and 1 second works well here.

Right-click on the Event to assign it to the BattlegroundBank, save your project, and Build the Banks before returning to Unity.

We could trigger the explosion sound effect in the DestroyRocket code, but let's use a similar approach to earlier:

- Select the Rocket prefab.
- Click on Add Component in the Inspector to add an FMOD Studio Event Emitter.
- Set the Play Event parameter to Object Destroy.
- Click on the magnifying glass icon to select the ExplosionSFX Event.

Play the game, and you'll hear your sound effects when the Rockets explode. You might want to adjust the Maximum Distance attenuation value in FMOD if you find the explosions get too quiet too quickly.

We're almost there – but it's still missing a couple of elements – there is no sound when the Rocket is fired, and it will help to add a rocket trail sound.

The rocket firing sound effect will always be heard at the same relative position to the player – it should be localized where we see the rocket launcher. This means that it makes sense to use a 2D Event, so go back to FMOD, and right-click on the Weapons folder to create a new 2D Event. Name this "RocketFire."

We don't need much variation to this sound effect (the weapon should sound pretty much the same every time it's fired), so we can get away with a Single Instrument here. Use the SFX_Weapon_RocketLauncher_Blast01 file.

Add a second Audio Track to the Event, and bring in the SFX_Weapon_RocketLauncher_Reload01 file in another Single Instrument. Drag this slightly later in the timeline, and your weapon now reloads after every shot.

If you select the Master Track for the Event, you can see the pan pot on the right-hand side (if this is missing, this is because the Event output is set to mono. At the right-hand side of the Deck is a meter labeled "out." Right-click on here, and changing this to Stereo gives you a pan pot). Position the sound effect to match the visuals – you may find it easier to use headphones to set this accurately.

Assign the RocketFire Event to the BattlegroundBank, and we're ready to start on the rocket trail sound.

Right-click on the Weapons folder to create a new 3D Event, and name it "RocketTrail." Again we'll be using a Single Instrument, but this time with the SFX_Weapon_Rocket_Trail_Loop01.wav file. You'll need this to loop, so right-click on the Instrument to add a New Loop Region.

117

For an extra element of realism, we're going to add a Doppler shift (a change in pitch due to the relative velocities of the sound source and listener – you can hear this when a car drives past with the classic "neeeeyowm" effect).

Select the Master Track, and at the right-hand side of the Deck you can see an option for Doppler. Turn this on, and the Scale parameter becomes available. It's not currently possible to preview this in FMOD, so leave it at 100% – you can always come back and adjust it later. For now, just add the RocketTrail Event to the BattlegroundBank.

We're also going to need laser sound effects for our Sentry Drones (see Chapter 15). These will be 3D Events, using the SFX_Weapon_Fire_Laser audio files. Create an Event called LaserBolt in the Enemy folder, and assign it to a new Bank named "SentryBank." Use the same techniques we covered in this chapter to make a laser bolt sound effect, before saving and building your Banks, and heading back to Unity.

We have a few options for how to trigger the RocketFire Event (for example, we could use an FMOD Studio Event Emitter set to "Object Start" attached to the Rocket prefab). Just so you get to see as many different techniques as possible, we'll do this with the ShootScript.

Navigate to the MyScripts folder and double-click on the ShootScript file to open it up in the IDE.

We need to add a line of code to the FireRocket function, so place the cursor after the semicolon on line 19 (rocketClone.velocity = transform.forward * speed;) and press [Return] to add an empty line underneath.

Now paste in the following line of code:

FMODUnity.RuntimeManager.PlayOneShot ("event:/Weapons/Rocket Fire");

118

```
 1 using System.Collections;
 2 using System.Collections.Generic;
 3 using UnityEngine;
 4
 5 public class ShootScript : MonoBehaviour
 6 {
 7     public Rigidbody rocket;
 8     public float speed = 40;
 9
10     // Use this for initialization
11     void Start()
12     {
13         Cursor.lockState = CursorLockMode.Locked;
14     }
15
16     void FireRocket()
17     {
18         Rigidbody rocketClone = (Rigidbody)Instantiate(rocket, transform.position, transform.rotation);
19         rocketClone.velocity = transform.forward * speed;
20         FMODUnity.RuntimeManager.PlayOneShot("event:/Weapons/RocketFire");
21     }
22
23     // Update is called once per frame
24     void Update()
25     {
26         if (Input.GetButtonDown("Fire1"))
27         {
28             FireRocket();
29         }
30     }
31 }
```

Figure 13.4 ShootScript 2

As you have probably already figured out, this line of code is used to trigger the RocketFire event whenever the FireEvent function is called.

Finally, we need to add the RocketTrail. We'll attach this as a component on the Rockets, so find and select the Rocket prefab in the Project Tab. Click on "Add Component" in the Inspector, and add another FMOD Studio Event Emitter.

Set Play Event to Object Start, and the Stop Event parameter to Object Destroy (this is necessary to stop the sound effect when the rocket is removed from the Scene).

Click on the magnifying glass to set the Event to RocketTrail, and try it out in the game.

You'll probably find that you can't hear the sound effect – the rocket moves away from the listener quite quickly, and the sound is masked by the RocketFire Event. Try experimenting with the Max Distance setting in FMOD to fix this – but don't forget to save and Build your Banks to push these changes to Unity.

If you play the game, you might have already spotted a bug. If the player is moving forwards when the weapon is fired, then they can collide with their own rocket, causing it to explode.

Obviously this isn't ideal. In Chapter 19 we'll see how we could avoid this by using Layers, but for now we'll work around this by moving the ShootFromHere GameObject slightly further in front of the launcher.

It's a bit boring to leave the rocket as a white ball. We could start by creating a new Material from scratch, but when we imported the Particle Systems, it brought in a number of suitable assets.

Select and lock the Inspector on the Rocket prefab.

Now click on "All Materials" in the Favorites sidebar of the Project Tab. You should now be able to see all the Materials included in the Unity project – including one called "Particle Firecloud." To add this to the Rocket Prefab, you have to drag the Material under the "Add Component" button – you'll see the cursor icon change to a curved arrow when you're in the right place.

Play around with the Tint Color, and you should be able to end up with something that looks far more interesting.

Creating Weapon Sound Effects

While it's quite straightforward to record foley and prop sounds at home, weapons can be a much greater challenge – even if you have access to a rocket launcher...

You can resort to sample libraries (this is quite common), but you won't always find the exact sound that you need.

It's possible to create your own very usable sound effects from items you have around the house. For example, the sound effects included with the book resources were all made in a recording studio.

The explosion effects are made up from multiple layers, including:

- Pink noise bursts
- Blowing into a microphone while mouthing explosion sounds
- Distortion
- Reverb
- More distortion
- More reverb

The weapon handling sounds were made from a combination of model guns, tools, and camera tripods.

The laser sounds (see Chapter 15) were made using a classic sci-fi sound effect technique: a contact mic was clamped to one end of a slinky spring. The spring was stretched out and hit with a pencil to create a "pew" sound.

The reason we think lasers should make this noise is due to Ben Burtt, sound designer extraordinaire. He discovered this sound effect by hitting a steel cable with a rock, and put it to memorable use in the original *Star Wars* film.

Our game is starting to take shape! In Chapter 14, we'll see how we can make the player's footsteps change according to the surface they are walking on.

121

14

Changing Footsteps

Learning Outcomes

By the end of this chapter, you will be able to:

• Configure FMOD and Unity so that the player's footstep sounds change when they move from one surface to another.

At the moment, our game only has one surface. Let's mix things up a little by bringing in the Fort asset.

You'll need to follow the same steps we used for our RocketLauncher:

• Import the Fort.fbx model to the MyModels folder.
• Extract the Textures.
• Extract the Materials.

The model is quite large, so we'll need to prepare somewhere to place it. Select the Terrain and choose the Paint Height tool. Shift-click to set your paint height, and click and drag on the Terrain to flatten out a space large enough for it to fit.

Now drag the model into the scene and play the game. You'll find that the Fort is intangible – you can walk through the walls and floor.

We could fix this by covering it with primitive colliders, but as this is quite a simple model, we'll use a Mesh. Select the Fort, click on Add Component, and find the Mesh Collider (it's under the Physics category).

This will automatically select the same mesh used by the model (if you were using a more complex model, you could use a simplified version of its mesh here).

Our Fort now has solid, textured floors and walls. One thing that still seems out of place is that our footsteps don't change when we walk on the metal floor.

There are many different techniques that can be used to change footsteps, from simple colliders to splatmaps (which analyse the relative amounts of the textures painted on the Terrain under the player). We're going to use Raycasting, as it's a technique that has multiple uses in game design.

A Raycast can be though of as a virtual line, which projects from a set origin point, and returns information about the different colliders that it encounters, including the Tag, the distance to, and the location of, the collider.

Before we get started coding, we'll need to create our metal footsteps. In FMOD:

- Right-click on the FootstepsGround Event and choose "Duplicate."
- Rename the duplicate "FootstepsMetal."
- Assign it to the Player Bank.
- Select the footsteps Multi Instrument, and click on "Clear" on the Playlist in the Deck.
- Go to File > Import Audio Files.
- Navigate to the FS_Metal_Boot_Walk files, and import them to the FMOD project.
- Drag the FS_Metal_Boot_Walk files from the Audio Bin into the Playlist.
- Save and Build the Banks, before returning to Unity.

We are going to change the First Person Controller script. Again you can either find the script in the Project Browser (the "All Scripts" Favorite search will be useful), or you can select the FPSController form the Hierarchy, and look for the component in the Inspector. Once you've found it, double-click on the script to open it up in your IDE.

123

We need to start by declaring our Raycast variable, so add the following code at line 43 (you'll find this code in the FMOD_FirstPersonControllerCodeForUnity text document):

```
40    private float m_StepCycle;
41    private float m_NextStep;
42    private bool m_Jumping;
43    private RaycastHit hit; // variable for Raycast
44
45    // Use this for initialization
46    private void Start()
```

Figure 14.1 Raycast code 1

We now have a Raycast type variable called "hit."

Now scroll down to the PlayFootStepAudio function (around line 160 – your line count may differ slightly), and change the code as follows (you'll find this code in the FMOD_FirstPersonControllerCodeForUnity text document):

Before

```
160    private void PlayFootStepAudio()
161    {
162        if (!m_CharacterController.isGrounded)
163        {
164            return;
165        }
166        FMODUnity.RuntimeManager.PlayOneShot ("event:/Player/Footsteps/FootstepsGround");
167    }
```

After

```
160    private void PlayFootStepAudio()
161    {
162        if (!m_CharacterController.isGrounded)
163        {
164            return;
165        }
166        if (Physics.Raycast (transform.position, Vector3.down, out hit)) {
167
168            if (hit.collider.tag == "SurfaceA") {
169                FMODUnity.RuntimeManager.PlayOneShot ("event:/Player/Footsteps/FootstepsGround");
170            }
171            if (hit.collider.tag == "SurfaceB") {
172                FMODUnity.RuntimeManager.PlayOneShot ("event:/Player/Footsteps/FootstepsMetal");
173            }
174        }
175    }
```

Figure 14.2 Raycast code 2

Raycast Footsteps Code

An "if" statement allows you to set conditions for when code is run. In this case, we have "nested" if statements.

- The first line (if (Physics.Raycast (transform.position, Vector3.down, out hit)) performs a Raycast.
- The Raycast originates at the same location as the player (transform.position), and points directly downwards (Vector3.down).
- The output results of this Raycast are given to the variable "hit" (out hit).

If the Raycast encounters anything, the code between the following curly brackets is run.

- If the hit Raycast encounters a GameObject with the tag "SurfaceA," then it plays the FootstepsGround FMOD Event.
- If the hit Raycast encounters a GameObject with the tag "SurfaceB," then it plays the FootstepsMetal FMOD Event.

The easiest way to follow this is by sketching out the logic:

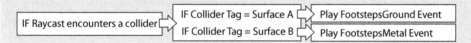

Figure 14.3 Nested "IF" statements

All that remains is to give different Tags to the Terrain and the Fort. Tags are reference words that you can assign to your GameObjects, which can be used to determine how they interact with code and other GameObjects.

- Go to Edit > Project Settings > Tags and Layers.
- In the Inspector, click on the disclosure arrow to display the Tags. There might already be some existing Tags (they are included with imported assets), and at the bottom right you'll see plus and minus buttons.
- Click on the plus symbol to create a new Tag.
- Name the Tag "SurfaceA," and click on Save.

- Repeat the process to create a SurfaceB Tag (note that it is important to get the spelling, letter case and spaces exactly the same as they're written in the code. SurfaceA will work, Surface A or surfaceA won't).
- Select the Terrain.
- At the top of the Inspector, you'll find the Tag parameter. Click here to change it to SurfaceA.
- Repeat the process for the Fort (but, of course, set this to SurfaceB).

Try this out in the game – your footsteps should automatically change as you move from one surface to the other.

It would be quite a simple process to add other surfaces to your game, and to modify the code and Tags accordingly.

Here's a challenge for you. Can you figure out how to change the game so that different Land Events are triggered, according to the different surfaces the player lands upon? Have a go at this for yourself – but back up your FMOD and Unity projects first, as it's quite easy to break the game when you first start playing with code!

Answer:

You could have taken several approaches to this – here's my solution.

- I started by creating a sub-folder in my FMOD Player folder called "PlayerImpacts."
- Inside here, I created two 2D FMOD Events called "LandGround" and "LandMetal," and assigned them to the Player Bank.
- I then used the FOL_Ground_Land_Impact and FOL_Metal_Land_Impact audio files to set up the Events, before saving and Building the Banks, and opening the FirstPersonController code in my IDE.
- Finally, I added the following code to the PlayLandingSound function (you can find this code in the LandingSurfaces text document in the book resources):

```
88    private void PlayLandingSound()
89    {
90        FMODUnity.RuntimeManager.PlayOneShot ("event:/Player/Vocals/Land");
91        if (Physics.Raycast (transform.position, Vector3.down, out hit)) {
92
93            if (hit.collider.tag == "SurfaceA") {
94                FMODUnity.RuntimeManager.PlayOneShot ("event:/Player/PlayerImpacts/LandGround");
95            }
96            if (hit.collider.tag == "SurfaceB") {
97                FMODUnity.RuntimeManager.PlayOneShot ("event:/Player/PlayerImpacts/LandMetal");
98            }
99        }
100   }
```

Figure 14.4 Play landing sound Raycast code

As you can see, this is pretty much identical to the code we used in the PlayFootStepAudio function – all that's been changed are the FMOD Events.

Don't be afraid to copy and re-use code – the classic quote is "Good coders code, great coders re-use" (there's another version of this that says "good programmers write good code; great programmers steal").

Raycasts have hundreds of uses in coding – and in Chapter 15, we'll see how they can be used to make our enemies shoot back at us…

127

The Enemies Shoot Back

Learning Outcomes

By the end of this chapter, you will be able to:

- Add a SentryDrone to your scene.
- Add a script to the SentryDrone so that it spins on the Y axis.
- Configure the SentryDrone so that it will lock on and fire at the player.

Though it's fun for a while to walk around shooting at Crates, after a while it becomes quite boring. In this chapter, we'll add some enemies to the game in the form of SentryDrones.

- The SentryDrones will spin on the Y axis.
- When the player is in front of the Drone, it will stop spinning and fire lasers at the player.
- The Drone will lock onto the player's position and continue to shoot, until the player moves out of sight for a set duration.

The SentryDrone 3D models are included in the book resources, so:

- Import the SentryDrone.fbx model to the MyModels folder.

- Extract the Textures.
- Extract the Materials.

Drag the SentryDrone model into the scene (somewhere close to the FPSController), and lift it about 1.5 units above the Terrain (this is easily done in the Inspector).

The SentryDrone will be intangible until we add a collider – a sphere collider should cover it quite nicely. It is now possible to walk into the Drone, and you can hit it with a Rocket.

That's great, but we need the SentryDrone to be destroyed when it's hit by the Rocket, so we need to create a new Script.

Navigate to the MyScripts folder, right-click in the gray-space of the Project Tab, and choose Create > C# Script. Name the script "DestroyMeWithRocket," and open it up in your IDE.

Copy in the following code (this code can be copied from the DestroyMeWithRocket1 text document).

```
1 using System.Collections;
2 using System.Collections.Generic;
3 using UnityEngine;
4
5 public class DestroyMeWithRocket : MonoBehaviour
6 {
7     // Use this for initialization
8     void Start()
9     {
10
11     }
12
13     void OnTriggerEnter(Collider other) //Code is called when a collider or rigidbody touches the trigger
14     {
15         if (other.gameObject.name == "Rocket(Clone)") //Determines tag of objects that will destroy the object
16         {
17             Destroy(gameObject); //Destroy the collided with GameObject
18         }
19     }
20 }
```

Figure 15.1 DestroyMeWithRocket 1

Save the script, return to Unity and drag the script onto the Sentry Drone.

DestroyMeWithRocket Script Walkthrough

Like the DestroyRocket code, this uses the OnTriggerEnter Method.

When the collider detects a collision with another gameObject, the script checks if the name of the other object is "Rocket(Clone)." If the name matches, it destroys the object that the script is attached to.

Play the game, and fire a few Rockets into the air. You can see each Rocket name appear in the inspector, and that they're named "Rocket(Clone)". This is because they are instantiated instances of the prefab, so the code has had to be been written to reflect this.

We could have used tags instead of the object name, but this would have involved creating another tag. If you have multiple types of weapon in your game, then this might be a better approach.

This has been a good start, but what good is a Sentry Drone that doesn't shoot back? Let's start by creating our laser bolt prefab.

- Select the SentryDrone in the Hierarchy, mouse over the Scene, and press [F] to focus the Scene.
- Go to GameObject > 3D Object > Capsule.
- Name the new GameObject "LaserBolt".
- This gives us the basis of our bolt, but it's going to need some changes. Start by ticking the "Is Trigger" parameter of the Capsule Collider component, so that we can use this to trigger code (just like we did for our Rocket).
- Scale the LaserBolt so that it looks as if it could come from the barrel of the SentryDrone (it will help to move and rotate the bolt so that you can see how it looks, but don't worry too much about the position transform values, as only the scale parameter will have any eventual effect).
- Apply a Rigidbody component.
- De-select the "Use Gravity" checkbox.

This is a good start, but the LaserBolt looks dull and white. Lasers should glow, so go to the MyMaterials folder, right-click in the gray-space and choose Create > Material. Name this "LaserGlow," and drag it onto the LaserBolt GameObject.

In the Inspector, you can click on the albedo box (next to the pipette icon) to set the base color.

To add a glow, check the "Emission" option, and click on the color box to set the emission color.

Finally, we'll need sound effects and explosions for the laser bolts. We can re-use some of our existing assets here. Let's start with the Explosions sound effect.

- Find and select the Rocket prefab in the Project Tab.
- Click on the cog icon at the top right of the Explosions FMOD Studio Emitter, and choose "Copy Component."
- Select the LaserBolt GameObject, and click on the cog for any of its existing Components (apart from the Material), and choose "Paste Component as New." You will now have the same explosion sound effects for your lasers and Rockets (though if you want to use different sound effects for each, there's nothing stopping you!).
- Hopefully you already created the LaserBolt FMOD Event in Chapter 13, so click on "Add Component" and add an FMOD Studio Event Emitter.
- Set the Play Event to Object Start and the Stop Event to Object Destroy.
- Click on the magnifying glass icon, and set the Event to LaserBolt.

All that's needed is the Explosion VFX. We'll re-use the DestroyRocket script, so navigate to the MyScripts folder, and drag the script onto the LaserBolt GameObject (remember that this will trigger the explosion effect when the laser collides with another GameObject, as well as destroying the bolt after a set time).

Once you've attached the script, you'll have to set the effect by dragging the Explosion prefab to the Explosion box in the Inspector component.

Our LaserBolt is all ready to go, so navigate to the MyPrefabs folder and drag the LaserBolt from the Hierarchy to the Project Tab (creating a new prefab).

We won't want the original LaserBolt in the scene, so right-click on this in the Hierarchy and select Delete.

That's the LaserBolt prepared, so now it's time to set up the SentryDrone.

We're going to need more scripts. We could have added lines of code to the existing DestroyMeWithRocket script, but keeping this separate means that we'll be able to re-use the scripts on other GameObjects. Different coders will take their own approach to this – some say that you should only have one script attached to a GameObject, others prefer to keep things separate and modular.

Return to the MyScripts folder, and create a new C# script called "SpinScript."

Copy in the following code (you can copy this code from the SpinScript text document in the book resources):

```
1 using System.Collections;
2 using System.Collections.Generic;
3 using UnityEngine;
4
5 public class SpinScript : MonoBehaviour
6 {
7     public float spinSpeed = 50; //Default spin speed - (Public means that it can be accessed from the inspector)
8
9     void Update() //This code is to be called every frame
10    {
11        transform.Rotate(Vector3.up, spinSpeed * Time.deltaTime);    //Rotate the GameObject around the Y axis by the speed value
12                                                                     //(Time.deltaTime makes this rotation speed independent of the game frame rate)
13    }
14 }
```

Figure 15.2 SpinScript code

SpinScript Walkthrough

Again, this is a very simple script. Every frame, it rotates the attached GameObject around the Y (up/down) axis.

The spinSpeed variable has been set to public, which means that it can be changed from the Unity Inspector (and that you will set it differently for each Sentry Drone in the game).

Add the SpinScript to the SentryDrone GameObject, and when you play the game, you'll see the Drone spin around. The spin speed can be adjusted in the Inspector, where you can fine-tune this to control the difficulty of the game. The faster the Drone spins, the quicker it will spot the player and start shooting. Negative speeds will make the Drone spin in the opposite direction.

Now to make the targeting and firing script. Create a new C# script in the MyScripts folder, name it "TargetingSystem," and open it in the IDE.

Copy in the following code, save the script and return to Unity (this is quite a long one, so it might be best to copy and paste the code from the TargetingSystem text document in the book resources. If you are typing in the code yourself, don't forget the letter c in acquire!).

```
1 using System.Collections;
2 using System.Collections.Generic;
3 using UnityEngine;
4
5 public class TargetingSystem : MonoBehaviour
6 {
7     private RaycastHit gotcha;
8     public Rigidbody laser;
9     public float laserSpeed = 40;
10     public float range = 40;
11     private float targetTime;
12     public float firingInterval = 2;
13     public float stopLookingAfter = 2;
14     private GameObject aimHere;
15     private bool acquireTarget;
16     //private FMOD.Studio.EventInstance sentryDroneAlert;
17
18     // Use this for initialization
19     void Start()
20     {
21         acquireTarget = false;
22         GetComponentInParent<SpinScript>().enabled = true;
23         aimHere = GameObject.FindWithTag("Player");
24     }
25     void FireLaser()
26     {
27         Rigidbody laserClone = (Rigidbody)Instantiate(laser, transform.position, transform.rotation * Quaternion.Euler(90, 0, 0));
28         laserClone.velocity = transform.forward * laserSpeed;
29     }
30     // Update is called once per frame
31     void Update()
32     {
33         targetTime += Time.deltaTime;
34         if (targetTime < firingInterval)
35         {
36             transform.parent.LookAt(aimHere.transform);
37         }
38         if (targetTime >= stopLookingAfter)
39         {
40             GetComponentInParent<SpinScript>().enabled = true;
41             acquireTarget = false;
42         }
43         if (Physics.Raycast(transform.position, transform.forward, out gotcha, range))
44         {
45             if (gotcha.collider.tag == "Player")
46             {
47                 GetComponentInParent<SpinScript>().enabled = false;
48                 if (acquireTarget == false)
49                 {
50                     //sentryDroneAlert = FMODUnity.RuntimeManager.CreateInstance ("event:/Enemy/SentryDroneAlert");
51                     //sentryDroneAlert.set3DAttributes(FMODUnity.RuntimeUtils.To3DAttributes(gameObject));
52                     //sentryDroneAlert.start();
53                     if (targetTime > firingInterval)
54                     {
55                         FireLaser();
56                         targetTime = 0;
57                         acquireTarget = true;
58                     }
59                 }
60             }
61         }
62     }
63 }
```

Figure 15.3 TargetingSystem code

- Select the SentryDrone in the Hierarchy, mouse over the Scene, and press [F] to focus on the Drone.
- Go to GameObject > Create Empty. The new GameObject will be placed at the same transform position as the SentryDrone.
- Name this "ShootLaserFromHere," and drag it on top of the SentryDrone (making it a child).

133

- Move the ShootLaserFromHere GameObject slightly in front of the SentryDrone.
- Drag the TargetSystem script onto the ShootLaserFromHere GameObject.

Now we have to configure the script component. Select the FireFromHere GameObject, and lock it to the Inspector. Drag the LaserBolt prefab into the Laser box to set what will be fired.

Finally, we need to set the player tag. Unlock the Inspector and select the FPSController. The Tag parameter is at the top of the Inspector, and you should find that there's already an option for "Player" (if not, select Add Tag at the bottom of the menu to go straight to the Tags and Layers menu).

TargetingSystem Code Walkthrough

Let's break this down section by section, starting with the declared variables.

- gotcha: a Raycast, used to detect the player in front of the SentryDrone.
- laser: this is used to define the fired object (in exactly the same way as the rocket Rigidbody was used in the ShootScript).
- laserSpeed: sets the speed of the fired laser.
- range: sets the Maximum Distance that the player can be detected by the Drone.
- targetTime: measures how long that the SentryDrone has had sight of the player.
- firingInterval: sets how often the SentryDrone fires.
- stopLookingAfter: sets how long after the SentryDrone loses sight of the player before it returns to its patrol.
- aimHere: sets the GameObject that the target looks for.
- acquireTarget: used to prevent the "Alert" sound from triggering continuously.
- sentryDroneAlert: an instance of an FMOD Event that will be called when the SentryDrone locks on to the player.

Start Method
Code inside the Start Method is run as soon as the script is enabled (in this case, when the game starts). It's used here to:
- Set the acquireTarget boolean to false.

- Turn on (set to "true") the SpinScript component on the GameObject's parent (i.e. the SentryDrone – remember that this script is placed on the ShootLaserFromHere GameObject).
- Sets the aimHere value to the GameObject tagged as "Player."

FireLaser Function

When this function is called, an instance of the laser Rigidbody is instantiated and fired forwards at the speed set by the laserSpeed variable. This is slightly different from the ShootScript – we have "transform.rotation * Quaternion.Euler(90,0,0))" at the end of line 27.

This is because when gameObjects are instantiated, the rotation of their prefabs doesn't have any effect. This would result in our laser bolts appearing to travel sideways, so we need to rotate them – and that's exactly what a Quaternion.Euler does.

Update Method

Code inside the Update Method is run every frame.

Time.deltaTime is the time taken to complete the last frame. This function is used whenever we need to count time – in this case the line "targetTime += Time.deltaTime;" adds this value to the targetTime variable.

The remainder of the code is a series of nested "if" functions. The easiest way to follow this logic is with a diagram:

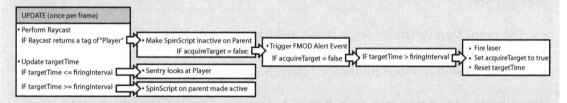

Figure 15.4 TargetingSystem Logic

1) If the targetTime value is less than the firingInterval value, then the parent (the SentryDrone) will look at (turn to face) the player.
2) If the targetTime value is greater than (or equal to) the stopLookingAfter value, then the SpinScript on the parent will be switched on, and the acquireTarget boolean set to false. This is used to reset the Drone after it loses sight of the player for a period of time (the firingInterval).
3) A Raycast is performed. If it returns a tag of "Player," then it disables the SpinScript on the parent.
4) If the Raycast returns a tag of "Player," if the acquireTarget boolean is false, then it will trigger the FMOD SentryDroneAlert Event (though this is currently commented out).

5) If the Raycast returns a tag of "Player," if the acquireTarget boolean is false, if the targetTime variable is greater than the firingInterval, then the FireLaser function is called, the targetTime value reset to zero, and the acquireTarget boolean set to true (this prevents the Alert sound effect from continuously triggering while the player is in front of the SentryDrone).

Note that lines 16 and 50 to 52 are in grayed-out italics. This is because they have been "commented out" – the two forward-slashes at the start indicate that the rest of a line is inactive. These are used to add notes to code – you'll have seen them used several times in scripts.

Commenting is good practice – it's very easy to forget exactly how (and why) you approached your code, and this will make it much easier to remember. Get into this habit early on, and trust me, you'll reap the benefits later on!

Commenting out code is also an easy way to make lines inactive without having to delete them. We haven't created an FMOD Alert Event yet, so we would get an error message if Unity tried to call it. Once we get round to making this, we can go back and delete the forward-slashes, re-activating the code.

As you can see, we have called the FMOD Event using a different technique than we used in the ShootScript. This is because the Alert Event will be 3D, so we need to assign a transform position.

- The DroneAlert variable is set so that it will trigger the Alert FMOD Event.
- The transform position of the SentryDrone is passed to the DroneAlert.
- The DroneAlert is played (DroneAlert.start();).

When you're writing your own code, it helps to sketch out the logic before you start, otherwise you'll find it very easy to get lost. Keep these sketches – they can be invaluable if you ever need to come back to your code at a later date!

So that just leaves the Alert sound effect:

- Return to FMOD.
- Create a 3D Event in the Enemy folder called "SentryDroneAlert" (use the SFX_Drone_Alert file in a Single Instrument).
- Assign the Alert Event to the SentryBank.
- Save and Build your FMOD Project.
- Open the TargetingSystem script in your IDE, and remove the forward-slashes on lines 16 and 50 to 52.
- Save the script and return to Unity.

Try this out. The SentryDrone should lock on and fire at the player. There's no repercussion to being hit yet – we'll get to that in a later chapter.

If you're finding that the lasers explode immediately, then try moving the FireLaserFromHere further in front of the SentryDrone – it might be catching on its collider.

In Chapter 16 we'll be putting the finishing touches to our SentryDrone, and looking at Parameter control in FMOD.

16

Parameters

Learning Outcomes

By the end of this chapter, you will be able to:

- Add a Scatterer Instrument to the scene.
- Make use of the Distance and EventConeAngle built-in FMOD and Unity parameters.

We're going to add a sound effect to the Drone – a humming sound, with occasional electric sparking. We started this sound effect way back in Chapter 5 – you should have an FMOD Event called "SentryDroneNoises." This gives us our looping Drone sound, so all that's needed is the sparking.

We could use a looping file. If it's long enough, you shouldn't notice any repetition. Of course, this will result in large audio files, which will bloat our game resources.

Instead, we're going to take this opportunity to look at Scatterer Instruments. These constantly spawn 2D or 3D sounds, and are great for creating generative effects.

Navigate to the SentryDroneNoises Event (it should be under the "Enemy" folder). It will help to tidy things up a little before we go any further, so:

- Double-click on the existing track name and rename it "Humm."
- Right-click here again to "Add Audio Track." Name this "Buzz."
- Right-click in the track to add a new Scatterer Instrument.
- Extend the edges of the Scatterer Instrument Trigger Region so that it fills the existing loop region.

We'll need to import our sparks, so go to File > Import Audio File, and find and open the SFX_Electric_Generator_Spark files. This will display the Audio Bin, so you can drag these files into the Playlist of the Scatterer Instrument in the Deck (note that the Playlist actually says "Add Instrument." This gives you some indication that Scatterer Instruments can be quite complex, with entire Events triggered. We won't be taking it this far in our game, but I'd recommend experimenting with this further for yourself).

Play the Event, and you'll hear spark sounds continuously triggered. With the Scatterer Instrument selected, you can adjust:

Min and Max Spawn Interval

The minimum and maximum times between triggered Instruments. The default settings are playing the sparks too frequently to my taste, so I'm setting this from around 800ms to about 3 seconds.

Min and Max Scatter Distance

This uses level variation to give the impression that Instruments are being triggered at different distances. That's not what we want here, so I'm going to set Min and Mix both to 1 meter. If I need any volume randomization, there's a dedicated "Vol Rnd" pot.

As you can probably guess, the Pitch Rnd control allows you to randomize pitch…

You might have noticed that even though the Playlist is set to randomized (the dice icon is yellow), the Scatterer Instrument always alternates between the two samples. This is because FMOD avoids triggering the same Instrument twice in succession – even when set to Random. There are several workarounds to this:

- Right-click on one of the Instruments in the Playlist and activate the "Set Play Percentage."

- Change the Play Percentage values. I've experimented with this, and have found that once I set the Instruments to 60%/40%, it will occasionally trigger the same Instrument twice in a row.
- Drag the same samples into the Playlist again. The duplicates will be treated as separate Instruments, and there will be a one in three chance of a repeat.

The spark sounds are randomly panning between the speakers, which, again, isn't what we need for our SentryDrones. Pan width isn't an option in the Scatterer Instrument, so we'll have to fix this some other way.

Select the Buzz track (the yellow box should cover the entire track), and look at the left-hand side of the Deck. The meters indicate that the track input is configured for 5.1 surround. Right-click on the meters, and you are given the option to change this to Mono.

The SentryDroneNoises Event is all good to go, so assign it to the SentryBank, Save and Build your FMOD Project, and return to Unity.

140

- Select the SentryDrone, and click on "Add Component" in the Inspector to add an FMOD Studio Event Emitter.
- Set the Play Event to "Object Start" and the Stop Event to "Object Destroy."
- Click on the magnifying glass icon to select the SentryDrone Event, before giving it a try in the game (you might want to turn off the Target Shoot script on the ShootLaserFromHere GameObject, so that you can hear the sound effect without getting shot at).

Parameters

Parameters are values controlled by the game states. Examples could include:

- The health of a player (we'll use this to add a filtering effect when the player's health drops below a certain value).
- The distance to an enemy (e.g. the music could change when you get close enough to an end-of-level boss).
- The speed of an object (e.g. the faster a vehicle goes, the higher the engine pitch gets).

We've already been using a Distance parameter with our 3D sounds – the further away the sound source, the quieter it gets. We can use this parameter to further enhance the impression of distance – if you get further from a sound emitter, you should hear less high-frequency content.

At the top of the SentryDroneNoises Event, next to the Tab labeled "Timeline" is a plus symbol. Click here, and choose "Add Parameter."

At the bottom of the pop-up box, click on "New Parameter." This displays the "Add Parameter" window. We'll start by setting the Parameter Type.

User Parameters are created by the game coders. We'll be creating some of these ourselves later, but for now, we'll use a Built-In Parameter. Built-In parameters are pre-configured – they are already passed through from the Game Engine, and require no additional coding. They include:

- Distance (distance to sound source from the listener).
- Direction (horizontal angle of the sound source from the listener).
- Elevation (vertical angle of the sound source from the listener).
- Event Cone Angle (horizontal angle of the listener from the sound source).
- Event Orientation (angle difference between the front of the listener and the front of the sound source).

Select "Distance," name the parameter "SentryDistance," set the min and max parameter range to match the Min and Max Distance values you've used for the SentryDrone (this will be 0 to 20 by default, but you might have changed this back in Chapter 2).

You are now in the SentryDistance Tab (you can change back to the Timeline at the top of the Event Editor). You've also added a pot in the Transport section of the Edit window called "SentryDistance," which you'll be able to use to preview the parameter control.

Select the Master Track, and click on the plus button on the left-hand side of the Deck. This allows you to add plugins and effects (you can get to the same menu options by right-clicking anywhere in the Deck that isn't a parameter control). Choose "Add Effect > Multiband EQ."

141

This adds a multi band EQ, complete with a low pass filter. I don't want the filter to have any effect yet, so raise the cut-off frequency up to the top (it's easiest to do this using the parameter box, so you don't change the Q/slope rate).

Right-click on the yellow parameter box, and choose "Add Automation."

This adds an automation lane to the Master Track. At the moment, there isn't any automation written, so the current parameter setting is shown as a dotted red line (if you click on the disclosure arrow on the right of the Multiband EQ, you can also show and hide the automation lane in the Deck).

Click on the dotted line to add a breakpoint towards the start of the automation lane (the line will go solid red). Now add a second breakpoint towards the end, and drag it down to about 500Hz.

You have now set the low pass filter cut-off frequency values for when you are at different distances from the SentryDrone.

Play the Event in FMOD, and use the SentryDistance pot at the top to hear how the timbre of the Event will change as the listener moves away from the sound source.

It will take a while to get this sounding natural. One thing that might help is to click and drag on the diamond between the breakpoints to change the automation line to a curve (and you can always add more breakpoints). Once you're happy with the results, Save and Build your FMOD project, and give it a try in the game.

Let's add another sound effect to the SentryDrone – one that indicates to the player that the Drone is looking in their direction.

Back in the FMOD SentryDroneNoises Event:

- Click on the plus symbol at the top of the Event Edit window to add another Parameter.
- Set this to Built In, Event Cone Angle, and name it "SentryCone" (the range for this parameter are fixed as 0 to 180).
- Go to File > Import, and bring in the SFX_Drone_Scan_01 audio file.
- Create a new Track in the SentryDroneNoises FMOD Event, and name this "Scan."
- Drag the SFX_Drone_Scan_01 file from the Audio Bin onto the track to create a new Single Instrument.

- Drag out the Single Instrument so that its trigger region goes from 0 to just past 45.
- Set the Single Instrument to loop (click on the loop icon in the Deck).
- Turn on the loop "Cut" button in the Deck.

This Instrument will trigger in a very different way than we've seen before. It will be played when the Parameter line is over the Trigger Region – in this case, whenever the listener is less than 45 degrees (approximately) from the front of the sound source.

The Cut button stops the loop as soon as the Parameter line leaves the Trigger Region (otherwise it will finish playing the current loop before stopping).

At the moment, the sound will stop very suddenly (the moment the Parameter line leaves the Trigger Region). Move your mouse to the top right of the Instrument, and you will see the cursor change to the fade-out tool. Use this to add a gradual fade to the effect (note that this won't just act as a fade out – it will also fade the Instrument back in when the Parameter value is lowered).

143

Save and Build the FMOD project, return to Unity and try out the game.

We'll need more than one Drone in our game, so navigate to the MyPrefabs folder in the Project Tab, and drag the SentryDrone into here from the Hierarchy.

You can now drag as many of these as you need into the Scene. You may find that they sink into the ground, but not to worry – you can select all of them at once in the Hierarchy, and then use the Move tool to lift them all the same distance above the Terrain.

It will also be worth placing all of the SentryDrones in an Empty GameObject. This will allow you to turn them all on and off at once, so that you can test out the game without being shot at.

Altering the individual spin speed, direction, and initial rotation of the SentryDrones will add some individual variation.

In our FPS game, there aren't many other opportunities to implement parameter controls. Later, we'll use the Score parameter to control the game music, but that's about all we'll be using. However, it's important for a game sound designer to understand these in more detail.

This is why I've included a bonus chapter (16b – Parameters 2). Here, you'll see how to use the speed of a GameObject as an FMOD parameter. There's no obvious place for this to be placed in the FPS level, so if you're in a hurry to get the game completed, you can skip this chapter, and come back to it later (however, there's some discussion of code included, which you might find useful, and we'll also be taking a look at FMOD's AHDSR Envelopes).

144

Bonus Chapter – Parameters 2

Learning Outcomes

By the end of this chapter, you will be able to:

- Use the velocity of a Unity GameObject as an FMOD parameter.
- Use the FMOD AHDSR Envelope to control the fade-out of an FMOD Event.

In this bonus chapter, we'll be taking a deeper look at parameter control.

We don't want to mess up our Battleground scene, so navigate to the MyScene folder, right-click in the Project Tab, and select Create > Scene. Give it a name – something along the lines of "AircraftScene" will do nicely. Now double-click on the scene icon to open it.

The Scene comes with a Directional Light and a Main Camera. We're going to be replacing the Main Camera, so right-click on this in the Hierarchy and delete it.

We need to bring in some more of Unity's Standard Assets, so go to Assets > Import Package > Vehicles. We only need the Aircraft, so to keep your project smaller, de-select the Car from the Import menu before clicking on Import.

Once that's done, head back to the same place, and select Assets > Import Package > Cameras. There's not too many assets here, so leave everything selected and click on Import.

The easiest way to find what we need from everything we just imported is to use the Favorites > "All Prefabs" at the top left of the Project Tab. Look for the "AircraftJet," and once you've found it, drag it anywhere you like in the scene.

Now we need to replace the camera we deleted earlier. Hunt for the MultipurposeCameraRig in All Prefabs, and drag it into the Scene.

With the MultipurposeCameraRig selected, you can see its settings in the Inspector. By default, the Auto Cam script should already have the AircraftJet(Transform) set as the Target, and the Auto Target Player option is ticked. Leave everything the way it is (if the Target isn't set to the AircraftJet, then drag it into this box from the Hierarchy).

The MultiPurposeCameraRig gives us a camera that can track and follow any GameObject. Place it anywhere in the scene, and it will fly to the Target and try to keep it in shot.

We're also going to need an FMOD listener, so click on the disclosure arrows in the Hierarchy to display and select the MultipurposeCameraRig > Pivot > MainCamera.

In the Inspector, click on Add Component, and add an FMOD Studio Listener.

The AircraftJet is a fantastic asset – unsurprisingly, it's a flying vehicle. Use the W and S keys to control pitch (up/down), and the A and S keys for roll (left/right). Holding the left mouse button will turn off the engine (if you want to use this in your own games, you can always replace the Aircraft model and keep the script assets).

If you'd left the Unity Audio on (see Chapter 10), you'd have noticed that the AircraftJet already makes sounds. To turn these off, select the AircraftJet, and turn off the Aeroplane Audio script in the Inspector.

We're going to use our own script, so navigate to the MyScripts folder, right-click in the Project Tab, and choose Create > C# Script. Name this "AircraftSounds," and double-click to open it in the IDE.

Copy in the following script (or paste it from the AircraftSounds text resource):

```
 1 using System.Collections;
 2 using System.Collections.Generic;
 3 using UnityEngine;
 4
 5 public class AircraftSounds : MonoBehaviour
 6 {
 7     private float aircraftSpeed;
 8     private FMOD.Studio.EventInstance aircraftEngine;
 9     private FMOD.Studio.ParameterInstance vehicleSpeed;
10
11     // Use this for initialization
12     void Start()
13     {
14         aircraftEngine = FMODUnity.RuntimeManager.CreateInstance("event:/Aircraft/AircraftEngine");
15         aircraftEngine.start();
16         aircraftEngine.getParameter("VehicleSpeed", out vehicleSpeed);
17     }
18
19     // Update is called once per frame
20     void Update()
21     {
22         aircraftSpeed = gameObject.GetComponent<Rigidbody>().velocity.magnitude;
23         //print (aircraftSpeed);
24         vehicleSpeed.setValue(aircraftSpeed);
25         if (Input.GetButtonDown("Fire1"))
26         {
27             aircraftEngine.stop(FMOD.Studio.STOP_MODE.ALLOWFADEOUT);
28         }
29         if (Input.GetButtonUp("Fire1"))
30         {
31             aircraftEngine.start();
32         }
33
34     }
35 }
```

Figure 16b.1 AircraftSounds script

AircraftSounds Script Walkthrough

The purpose of this script is to trigger an FMOD Event, to measure the speed of the aircraft, and to pass this value along to FMOD as a parameter. Before I go through this in detail, it will help to clarify one of the Unity coding conventions.

Variables

Variables should always start with a lower-case letter – for example, this script has a variable called vehicleSpeed.

However, you can see at line 16, VehicleSpeed is written with both upper-case and lower-case V. The upper-case V refers to the FMOD parameter name, and the lower-case v refers to the Unity script variable (in fact, this line tells Unity to send the vehicleSpeed variable value to the FMOD VehicleSpeed parameter).

The FMOD parameter name actually could have been written with an upper- or lower-case letter, and the code would still work (in fact, it could be given a totally different name). To make the code easier to follow, the same name is used for the Unity variable and the FMOD parameter, and the first letter case distinguishes between them.

This script only has three variables to declare:

- aircraftSpeed: obviously, the speed of the aircraft.
- aircraftEngine: a Unity script variable that will trigger an FMOD Event.
- vehicleSpeed: the Unity script variable that will send a parameter value to FMOD.

In the Start method:

- We set the aircraftEngine Event Instance so that it triggers an FMOD Event called "AircraftEngine."
- We start the Instance.
- We set the FMOD VehicleSpeed parameter so that it is sent the vehicleSpeed value.

In the Update method:

- We get the velocity of the gameObject that the script is attached to, and assign this to the aircraftSpeed variable.
- We pass the aircraftSpeed value to the vehicleSpeed variable.
- We check if the left mouse button is held down (Fire1). If it is, the aircraftEngine instance is stopped.
- We check if the left mouse button is released (Fire1). If it is, the aircraftEngine instance is started again.

Save the script, return to Unity, and drag the script onto the AircraftJet GameObject.

Next, we'll have to create our FMOD Event. In FMOD, create a new folder called "Aircraft." Inside here, create a new 2D Event called "AircraftEngine."

Now click on the plus symbol at the top of the Event Edit window to add a new parameter. This time, we'll need to leave it set to

User Parameter. Name this "VehicleSpeed." Don't worry about the Minimum and Maximum settings for now – leave it as 0 to 1, and we'll come back to this later.

There are several ways that we can configure a vehicle sound so that it responds to a speed parameter. We'll look at two of these:

Technique 1: Pitch Automation

Import the SFX_Aircraft_Engine_Low and SFX_Aircraft_Engine_High audio files to your FMOD project.

We're not going to be placing the Instruments in the Timeline – instead, we'll be putting them in the VehicleSpeed, so select the appropriate Tab at the top of the Edit Window.

Drag the SFX_Aircraft_Engine_Low into the Audio Track, to create a Single Instrument. Extend the Event Trigger region so that it covers from 0 to 0.6, and set it to loop in the Deck.

Next, drag the SFX_Aircraft_Engine_High into the same track, creating another Single Instrument. Drag out and position the Event Trigger to cover from 0.4 to 1.0, and set it to loop in the Deck.

Preview the Event in FMOD, and experiment with the VehicleSpeed parameter.

You can see a cross-fade where the Instruments overlap, where both Instruments will play together.

Select each Instrument in turn, and right-click on the Pitch pot in the deck to add automation lanes.

You can now set how the pitch of each Instrument will be affected by the parameter.

With quite a lot of fine-tuning, you should be able to get the two Instrument sounds to blend together, giving a smooth transition as you change the VehicleSpeed parameter. You'll need to experiment with the cross-fade (longer cross-fades will give much better results), automation breakpoints and curves, as well as the Instrument levels.

This is the "old" way of doing things in FMOD, but it can give excellent results if you put the time in (and if the vehicle audio assets are created/recorded with this end configuration in mind).

149

Technique 2: Autopitch

If you've tried out the first method, you might have noticed that changing the Event Trigger and cross-fades means you then have to adjust the automation.

This can make setting up smooth transitions between Instruments quite difficult. This is why Autopitch was introduced.

If you've already added pitch automation, get rid of it by right-clicking in the lane and choosing "Remove Automation."

Then select each Instrument in turn, right-click on their Pitch pots in the Deck, and choose Add Modulation > Autopitch.

This gives you two parameters: Root Pitch and Min Pitch.

Root pitch is the parameter value when the sound will play at its original pitch, so if you set this to 0.5, set the VehicleSpeed parameter to 0.5, and hit Play, you'll hear the sample played without any pitch shift.

Change the VehicleSpeed parameter, to hear the pitch change.

Min Pitch sets the lowest that the pitch will go when the parameter is all the way at the bottom of the Event Trigger. If this is set to 0.5, it will halve in pitch (i.e. go an octave lower).

The pitch shift scales according to the distance between the Min Pitch and the Root Pitch, so if a parameter change from 0.5 to 0 causes the frequency to halve, a change from 0.5 to 1 will increase the pitch by the same frequency (not pitch) amount (i.e. to a musical 5th above the root).

Once you've finished with the AircraftEngine Event, assign it to a Bank (I suggest a new Bank called "AircraftBank"), Save and Build your project, and return to the AircraftSounds script.

There's a commented-out section at line 23: //print (aircraftSpeed);

Make this line active by removing the forward-slashes, save the script and head over to Unity.

Print is available because our script derives from the MonoBehaviour class. It is used to display a message in the Unity Console. In this case here, it will show the aircraftSpeed value every frame.

Play the game, and you'll now see the aircraftSpeed updating at the bottom left of Unity (you can also see this by opening the Console Tab).

Wait a little while, and you should see that it peaks at around 190. This gives us our parameter range. Head back to your IDE, and put the forward-slashes back to re-comment out line 23.

Save your script, and return to FMOD. Right-click on the VehicleSpeed Tab to edit the parameter.

Set the Maximum value to 190, and click on OK. It will then offer you the options of Cancel, Don't Scale, and Scale. Choose "Scale," and FMOD will scale your previous settings (including the Root and Min Pitch values) to the new range.

Save and Build the FMOD project, and you should be able to try out your aircraft sound in the game.

Hopefully, everything's working. However, you'll probably find that the AircraftEngine stops very suddenly when the right mouse button is pressed.

Line 27 of the AircraftScript says "aircraftEngine.stop(FMOD. Studio.STOP_MODE.ALLOWFADEOUT);."

This means that we have the ability to control the way that the Event Instance stops.

Open the AircraftEngine Event in FMOD, and select the Master Track.

Right-click on the Volume pot in the Deck, and choose Add Modulation > AHDSR.

This has added an envelope to the Event – similar to one that you'd find on a synthesizer.

- A: Attack. The time taken to reach the Hold level.
- H: Hold. The time that the level is held for, before the Decay starts.
- D: Decay. The time taken to change from the Hold level to the Sustain level.
- S: Sustain. The level at which the signal is held at until the Release is triggered.
- R: Release. The time taken for the Event to fade out. Triggered when the Event is Stopped (if the Stop mode is set to Allow Fadeout).

There are a couple of other parameters, which aren't part of the acronym, but are still important:

- Initial: the level that the Attack starts from.
- Peak: the Hold level.

By default, adding the AHDSR has set a 1 second Attack and Release envelope. This is about right for what we need – but feel free to experiment with these settings before saving, building, and trying out the sound again in the game.

Creating the Aircraft Sounds

Most of you won't have access to jet aircraft, so you'll have to think outside the box for this.

The samples provided with the book resources were created in the studio, and are made up of layers including:

- A fan
- A hairdryer
- A hot-air gun/paint stripper
- A dehumidifier
- Filtered noise

You've probably got most of these items, so try recording and using your own sounds. You might need to make use of pop-filters and wind jammers, but if you don't have access to these, it's possible to put together a make-shift wind jammer using a sponge and a sock – there are tutorials that show you how to do this on YouTube.

In Chapter 17, we'll be returning to our FPS level, so save the scene, navigate to the MyScenes folder, and double-click on the Battleground scene.

Signal Routing in FMOD

Learning Outcomes

By the end of this chapter, you will be able to:

- Route Audio Tracks within an FMOD Event.
- Create and assign FMOD Events to Group buses.
- Use auxiliary routing to add effects to FMOD Events and Audio Tracks.
- Configure custom mixer views in the FMOD Mixer window.

Much like a DAW or mixing desk, FMOD has sophisticated signal routing capabilities, and when it comes to adding effects, mixing, and automation, it is essential we make use of these.

Audio can be routed within an Effect as well as within the Mixer. Let's start by looking at routing within an Event. The SentryDroneNoises Event should do nicely, so select it in FMOD and make sure you're in the Timeline Tab.

This Event has a low pass filter applied to the Master Track, so turn down the SentryDistance parameter pot to remove its effect (alternatively, select the Master Track, right-click on the Multiband EQ in the Deck, and select "Bypass").

As well as the Master Track, we have three Audio Tracks (Humm, Buzz, and Scan). Currently, all three Audio Tracks feed the Master Track.

Audio Tracks

As well as holding Instruments, Audio Tracks can be used to sub-mix within an Event. Right-click on any of the track names, and choose "Add Audio Track." Name the new track "SumBuzzHumm."

Select both the Humm and Buzz tracks (hold [cmd] (Mac) / [ctrl] (Windows) to make multiple selections), then right-click again, and change the outputs to SumBuzzHumm.

You can now use the SumBuzzHumm track to control the level, panning and effects of both the Humm and Buzz sounds (without affecting the Scan).

Return Tracks

These allow for parallel signal routing. They cannot hold Instruments – they get their signals from Sends.

- Right-click on a track name, and choose "Add Return Track."
- Name this "DemoReverb".

While you have the DemoReverb track selected, let's set up its effect.

- Click on the plus symbol to the left of the Volume pot, and choose Add Effect > FMOD Reverb.
- Select the Buzz track and click on the plus symbol to the right of the Volume pot in the Deck.
- Choose Add Send > To Event > DemoReverb.

You now have a Send pot in the Deck, which allows you to send a controllable amount of the Buzz track to the DemoReverb.

Pre/Post Sends

The position of the Send is "Post Fader," i.e. it comes after the Volume pot. This means that if you turn down the level of the Buzz, you will also affect the signal level going to the Reverb. This is usually the best way to set up reverb routing, as the wet/dry (reverb/no reverb) balance will not be affected when you change the track levels. However, if you ever have a reason to alter this, you can simply click and drag to re-arrange the order of Components in the Deck (this is actually one of my favorite features of FMOD – the ability to configure the signal path order allows you to set up your Sends with or without effects applied. I wish this was available in more DAWs!).

As the routing is parallel, it would be possible to add more effects to the DemoReverb track, without affecting the original "dry" signal. You can also add a Send to the DemoReverb from both Scan and Humm tracks, so that the same effect instance can be used on multiple sources. This can significantly reduce the signal processing load.

As you might have already guessed from the name, we're not actu-ally going to use the DemoReverb track, so right-click on the track name and delete it (this will also remove the associated Sends).

Reverb is a real resource hog, so instead of placing plugins on every track, instead, it is preferable to use a single reverb for all of your Events. Not only will this reduce the reverb signal processing load on the game, but it will ensure that all sounds will be placed in the same acoustic environment.

The human ear can ascertain an astonishing amount of information from reverb, from the size (and, to a certain extent, shape) of a room, to the absorptive properties of its surfaces. If it hears conflicting infor-mation from multiple reverb sources, it gives up, and just interprets is as "reverby." Using a single reverb instance maintains consistency, which helps the acoustic immersion of the player.

FMOD projects come pre-configured for this, with a Reverb Return that can be used by all Events and Tracks. We'll see this shortly when we take a look at the Mixer window, but you might already have spotted it when setting up your Sends (there is an option of Insert Send > To Mixer > Reverb).

Adding Effects – Real-Time vs Pre-Rendered

We've already seen how to add effects to tracks, and how to change their order. But before we start putting effects on everything, let's con-sider the hows and whys.

All real-time effects take up processing power – power that could be used elsewhere in the game – and remember, graphics almost have priority over sound! So the first consideration should be whether to use live FMOD effects, or whether to use pre-rendered/printed/baked audio files (i.e. to apply the effects in your DAW, and include them in the bounced audio files).

Playing an audio file with the effects already applied takes the same processing power as playing it without.

Printed effects obviously can't be controlled by real-time parameters, so they won't be suitable all of the time. Also, if you need sections of the game to have the sound with the effect and other sections without, then you will need twice as many audio files.

This makes your game file size bigger, and also affects memory allocation – if you need to frequently switch between sounds, both audio files will need to be held ready for use in RAM.

In Chapter 28, we'll look at optimizing our game, and see how to use the FMOD Profiler to measure and test its performance. For now, just be aware of the available options.

We'll be sticking to the stock FMOD effects in this book, but you'll find third-party effects under the "Plug-in Effects" sub-menu, including offerings from McDSP and Google. These are worth experimenting with, but some cannot be included in a Build unless you pay for a license.

The FMOD Mixer

In FMOD, go to Window > Mixer. At the moment, it will look quite empty. At the top, you'll see that it's set to Mixing Desk. This shows you all of the Groups and Return channel strips that are in your game, and at the moment we only have the default Reverb Return.

(Note – it's quite easy to accidentally change the Mixer window to "compact" narrow strips. Toggle this on and off with the [C] key.)

FMOD is designed so that you should use Groups for your Events. These are sub-mixes, which allow you to control multiple events at once, as well as to set up automation snapshots and sidechain effects. When you consider the number of Events that can go into a game, you'll understand why (for example, if you want to turn down the level of all the sounds the player can make, it will help to be able to do this with a single fader).

Let's start by grouping the two Footsteps Events. The easiest way to do this is:

- Select both the FootstepsGround and FootstepsMetal Events in the Routing Tab on the left of the Mixer window.
- Right-click, and choose "Reroute into New Group."
- Name the Group "Footsteps."

You can now see the Group in both the Routing Tab and the Mixing Desk.

Repeat the process to place the player vocalizations (the Jump and Land/Impact Events) into a Group called "Vox."

We can also group our groups: Right-click in the gray-space of the Routing Tab, and choose "New Group." Name this "Player," and drag the Vox and Footsteps on top to place them in the Group.

It helps to sketch out what grouping you'll need in your project before you go much further here. Do you want your weapon firing Events to be included in the Player Group, or will you need a separate Group for all weapon sounds, including those made by the SentryDrone? (if necessary, you can have Groups that only contain a single Event). Here's my current plan, and I recommend that you use this for your own routing.

157

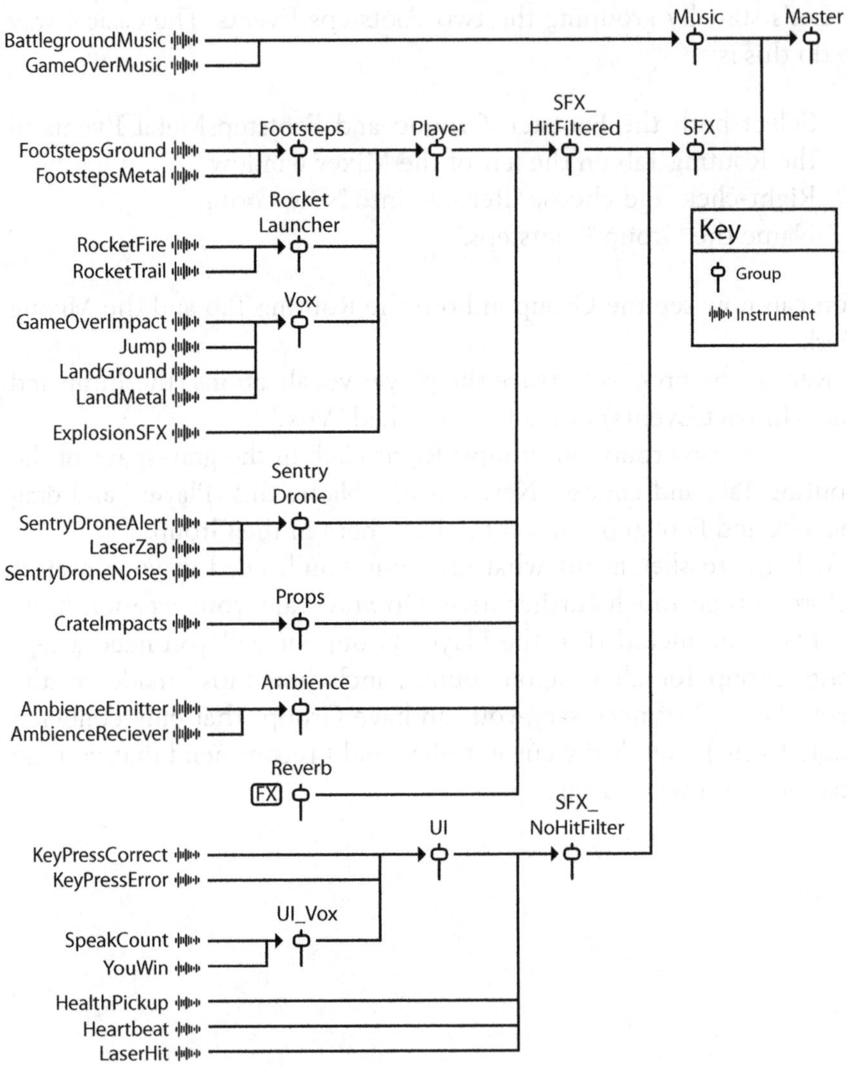

Figure 17.1 Signal routing plan

Note that I've summed all of the sound effects into one Group called "SFX_HitFiltered," and prepared another called "SFX_NoHitFilter." This will be used later on, so that a low-pass filter effect can be added to some elements of the sound effects (this will be applied when the player is hit by a laser – see Chapter 21).

I've also created Groups ready for when we add our music and UI sounds (see Chapters 24 and 25).

Mixer Views

When you're mixing the game, you'll need access to both Groups and their constituent Events, without having to go back to the Event Editor every time.

If you select an Event in the Routing Tab, you'll see the Deck shows the Master Track settings, so that's a start.

At the top of the Mixer window, change to "Selected Buses." Now the window shows faders for whatever you select in the Routing Tab (you can show multiple Events by holding [shift] or [cmd] (Mac) / [ctrl] (Windows) to add to your selection).

That can be helpful, but we're going to set up custom Mixer Views.

Click on the Plus symbol at the top of the Mixer window (next to where it says "Selected Buses") to add a Mixer View. Name this "Player." As the screen tells you, drag in Groups and Events from the Routing Tab to add them to the View.

Now you can view all of the Player Groups and Events in one place. Groups and Events can appear in multiple Mixer Views, so it's worth adding the Reverb Return here as well.

You can spend some time making a number of Mixer Views, but I tend to create them as and when they are needed.

Next, let's see how we can configure our Reverb routing. We could set up Sends from individual Events, but later on we'll use Snapshot automation to turn on and off our reverb when we go in and out of the Fort – and snapshots only work on Groups.

Select the Player Group to see its settings in the Deck. Click on the plus symbol to the right of the Volume pot to add a post fader Send to "SFX_HitFiltered > Reverb" (remember that we routed the Reverb via the SFX_HitFiltered group earlier in this chapter). You will be able to see this added to the channel strip (you could also have right-clicked on the channel strip to add your Send).

Turn up this pot, and all sounds in the Player Group will have reverb added.

VCAs

VCA (Voltage Controlled Amplifier) faders are found on some analogue mixing consoles to control multiple signals at once, without having to bus them together – and that's exactly how we use them in FMOD.

I configured FMOD so that the player weapon-firing Events (RocketFire and RocketTrail) were in the Player Group, and the LaserBolt Event was in the SentryDrone Group. But what if I want to control the level of all my weapon-firing sounds at once? You guessed it, by using a VCA Group.

Select the RocketFire, RocketTrail, and LaserBolt Events in the Routing Tab, then right-click and choose Assign to VCA > New VCA. Name this "WeaponsFire."

You won't see VCAs in the Routing Tab – you'll have to change to the VCA Tab. However, you can still drag them from here into your Mixer Views. They'll appear with a red fader cap.

As the signals are never actually summed, you cannot apply effects and Sends to VCAs, but they still provide a very useful mixing option.

One quick note – if you're using post-fader Sends, don't put your Returns into the same VCA as the dry signals, otherwise the VCA will affect the wet/dry ratio (e.g. turning down the VCA will affect the signal going to the effect, as well as the signal coming back).

In Chapter 18 we'll take a look at FMOD's effects plugins. So Save and Build your FMOD project, and I'll see you over the page.

FMOD Effect Plugins

Learning Outcomes

By the end of this chapter, you will be able to:

• Configure FMOD's effects plugins.
• Use FMOD plugin presets.

FMOD comes with a number of plugins, but even if you're already familiar with effects, it's worth taking a closer look at these – they don't all behave quite in the way that you might expect.

Reverb

Reverb is caused by sound reflecting off different surfaces in the environment, before arriving at the listener. Understanding how these reflections behave is the key to making sure your reverb sounds right.

FMOD Reverb Plugin Parameters

Reverb Time
This is the time it takes the reverb to decay by 60dB (it's sometimes referred to as "RT60"). While this may have the most obvious effect,

it's not the only parameter to get right – after all, a tiled bathroom could have a similar reverb time to a cathedral.

Early Reflections and Late Reverberation

The FMOD Reverb plugin divides the reverb into two sections: Early Reflections and Late Reverberation.

Early Reflections have only bounced off one or two surfaces, and tell you about the size and relative dimensions of the environment.

The Early Delay parameter sets the time (in milliseconds) before the first reflection is heard. Setting a longer value can give the impression of a larger room.

Late Reverberation is made up of multiple reflections, which arrive too closely together to distinguish between them. They tell us about the absorptive properties of the materials in the room.

The Late Delay parameter sets the time after the Early Delay (in milliseconds) before the first late reverberation arrives.

162

HF Reference

Porous materials absorb high frequencies (it's far more difficult to absorb low frequencies), so they will decay much quicker. The HF ratio parameter controls the decay rate ratios between high and low frequencies – at 100%, they will decay at the same rate. At lower values, the high frequencies will decay more quickly.

HF Reference sets the crossover frequency between high and low frequency ranges.

Density

Density sets the spacing of the Late Reverberations, and controls the tonal effect that reverb has. In a symmetrical room, reflections between parallel surfaces result in "room modes" that give a resonant ring (this is more pronounced in small spaces, where it is sometimes called the "small room problem"). The lower this parameter is set, the more pronounced the resonance effect – great for giving the impression of a small space.

Diffusion

Diffusion sets the spacing between the reflections. The larger the room, the lower you should set the Diffusion (though turning this down too much can give "flutter echoes").

Low Freq, Low Gain, and High Cut

These are the controls for a shelving filter applied to the reverb.

High Cut sets the cut-off frequency for a low pass filter.

Early/Late

This pot allows you to control the level ratios between the Early Reflections and the Late Reverberation.

Wet/Dry

That just leaves the Wet and Dry levels, which control the amount of the effect. If the effect is configured as a Return, this should be left at 100% wet, 0% dry (the dry signals can be heard in their original Events).

163

Convolution Reverb

FMOD also includes a convolution reverb. These are based on impulse responses (IR) – recordings of an acoustic environment when an impulse is played (a "true" impulse can't actually exist in real life, but we can approximate one – for example, by firing a starter pistol or popping a balloon).

Convolution reverbs can give very realistic results, but they are processor intensive, and can't easily be adjusted – essentially, the IR files are reverb presets.

If you'd like to experiment with this, I've included two files with the book resources (these were recorded in my studio before any acoustic treatment was installed). Simply drag one of them onto a Convolution Reverb in the Deck.

The only other parameters available in the Convolution Reverb are the wet and dry signal level controls.

EQ

EQ, or "Equalization," is the level control of different parts of the frequency spectrum.

FMOD Multiband EQ

FMOD's Multiband EQ has five bands, labeled A to E, which can each be configured as a different type.

Pass Filters

First we have LP (Low Pass) and HP (High Pass) filters. An LP Filter allows low frequencies to pass, and as you might expect, an HP filter allows high frequencies to pass. There are several Pass filter "slope rate" options (12dB, 24dB and 48dB per octave), and we'll see what the differences between these are in a moment.

The Pass filters have two parameters:

Frequency

This sets the position of the filters (strictly speaking, the cut-off frequency is actually attenuated by 3dB, but we don't need to go into this level of detail here).

On an LP filter, frequencies lower than this are unaffected.

On an HP filter, frequencies higher than this are unaffected.

The higher (or steeper) the slope rate, the more immediate the effect of the filter. For example, a LP filter set to 12dB (per octave) will only slightly attenuate frequencies just above the cut-off frequency. If this is changed to 48dB (per octave), they will be attenuated far more significantly.

Q

On the Pass filter settings, this is actually the Resonance control. This gives a spike in level just before the filter roll-off, which can be used to emphasize the filter position.

Shelving

A Shelving EQ applies a roughly uniform amount of gain to signals to one side of the "Shelving Frequency." Unlike Pass filters, they can boost and cut using the "Gain" parameter.

- A Lowshelf affects signals lower than the Frequency parameter setting.
- A Highshelf affects signals higher than the Frequency parameter setting.
- The Q parameter is disabled.

Peaking

This gives a "parameteric" band, which has an effect on a bell-shaped range of frequencies. These have three parameters.

Frequency

This control adjusts the center frequency. This is where the EQ will have the most effect.

Gain

This adjusts how much the center frequency is affected.

Q

Controls the bandwidth – how far either side of the center frequency is affected.

Bypass

I've missed out possibly the most important control of any plugin – the Bypass.

Once you start to EQ anything, your ears become attuned to the changes you've made. It's quite easy to get a little lost, and end up with a sound that's worse than when you started! Make sure you A/B your EQ, and regularly refer back to the original.

3EQ

This splits the signal into three frequency bands. Parameters include:

- Low, Mid and High (these are gain controls, which allow you to set the level of each band).

165

- X-Low and X-High (these set the crossover frequencies, which determine which frequencies are included in each band).
- X-Slope (this allows you to change the slope rate of the crossover filters [in dB per octave]. The lower this value, the more overlap there will be between the bands).

Channel Mix

"Channel" here refers to the panning of a signal – for example, placing one of these on a stereo track will allow you to independently adjust the left and right channels.

To see exactly what this plugin does, select the Master Track of an Event, right-click on the "In" meter at the right-hand side of the Deck, and change to 5.1 surround, before adding a Channel Mix plugin. You will see six faders available – one for each of the channels. These are labeled 0 to 5, which isn't too helpful. Change the Grouping to 5.1, and you will see the labeling changes to show FL (Front Left), FR (Front Right), C (Center), LFE (Low Frequency Effects), SL (Surround Left) and SR (Surround Right).

Once you've played around with this, don't forget to change the Input back again!

Chorus

Chorus is an effect used to thicken up a sound, by blending it with a slightly pitch-shifted copy. Parameters include:

- Rate (the frequency at which the pitch is modulated).
- Depth (how much the pitch is modulated).
- Mix (how much of the pitch modulated signal is added).

If you've used a Chorus plugin on a DAW before, you might find that FMOD's version is a little different – for example, it modulates its pitch-shift by speeding up, then slowing down the sample playback. In addition, the Depth is not set in semitones, but instead this interacts with the Rate – a greater amount of pitch shift is achieved with a higher Rate setting.

Though this might not be great for music, it has fantastic potential in sound design – you easily can get some very unusual (but very usable) results.

Compressors

First thing to get straight is that we're not talking about data compression – we're looking at dynamic range compression.

Dynamic range is the difference between the quiet and loud signal levels. A compressor changes this – for example, it can be used to reduce or (if you know what you're doing) exaggerate the dynamics.

Parameters include:

Threshold

A compressor starts to have an effect when signals get higher than the threshold – if it is this all the way up, the plugin will have little to no effect.

Ratio

The basic function of a compressor is to reduce dynamic range, and the ratio sets how much dynamic reduction is applied – set this to 2:1, and it will reduce the dynamics above the threshold by a half.

To give you an example, if a signal comes in 8dB higher than the threshold, it will leave the compressor only 4dB higher than the threshold.

Attack and Release

You can think of a compressor as containing a small man with amazingly fast reaction times, and a fader that controls the signal level.

When he hears the signal level get louder, he pulls down the fader. Once it drops below the threshold again, he pushes the fader back up.

The Attack and Release are his reaction times – Attack is how fast he'll pull down the fader, and Release is the speed he puts it back up again.

These controls change compressors from simple level control devices, into Instruments that can affect the envelope of a sound.

For example, if you set a slow Attack time, you can allow the beginning of a sound to get past the level control, before the compressor has the chance to react to it. This is a great way to add a "bite" to the start of a sound.

Gain

As a compressor works by turning down the signal level, this would result in everything coming out too quiet. To compensate for this, we have the make-up gain, which turns the entire signal back up again.

Link

Unless the channels are linked, when you compress a stereo signal you will get panning position shifts.

Consider a stereo recording of three people: Dave, Mathew, and Ian. Dave is panned to the left (i.e. he is only in the left channel), Ian is panned to the right (i.e. he is only in the right channel) and Mathew is panned to the center (i.e. he is heard equally from both left and right channels).

If Ian starts screaming, this causes the compressor to gain reduce the right channel.

If the right channel is gain reduced, then the stereo position of Mathew will move to the left.

To avoid this, we need to ensure that if xdB of gain reduction is applied to the right channel, the same amount must be applied to the left (and vice versa). This is the function of the Link control.

Sidechain

This allows the compression to be controlled by another signal. To return to our easier analogy, think of this as giving the small man in the compressor a pair of headphones. The Sidechain allows you to unplug his headphones, then patch them to listen to a different signal. We'll be coming back to this in Chapter 27.

Limiter

A common use of a compressor is for "Limiting." This is simply a compressor with a very high ratio, and very fast attack time. They are used to stop signals getting too loud – for example, you could apply a limiter to affect all the explosion sounds, preventing them from deafening the player.

Though you could use the FMOD Compressor, it also comes with a dedicated limiter.

Parameters include:

Ceiling

This is the compressor's threshold. Limiting is considered to be infinity:1 (i.e. the output never exceeds the threshold). This means that if this is set to -3dB, there will be 3dB of "headroom" (before the signal clipping level).

Input

This allows you to adjust the signal level before the limiting stage.

Release

See compressor parameter.

Link

See compressor parameter.

169

Distortion

Does exactly what it says. The higher your set the Level, the more distortion you get.

Delay

A Delay works by feeding a copy of the signal into memory, then reading it out again at a later point in time. Parameters include:

Delay

The time before the signal is read out again. The FMOD delay plugin allows you to set this as high as 5 seconds, but be careful here – the longer the delay time, the more RAM the effect will use up.

Feedback

To get multiple repeats you could use more Delays (this would be known as a multi-tap delay). Instead, we can take a copy of the Delay output, and feed it back to the input, in what's called a "feedback loop." However, this would result in infinite repeats. This might sound quite interesting, but will quickly become very messy. Instead, we turn down the level of the feedback loop so that the repeat becomes quieter each time – and that's the function of the Feedback control.

This can be set from 0% (one repeat) to 100% (infinite repeats).

Wet/Dry Level

This allows you to set the levels of the dry (unaffected) and wet (delayed) signals.

Flanger

This is a similar type of effect to Chorus, but the FMOD Flange works using a modulated delay, which causes the output pitch to change. Parameters include:

Rate

How many times per second the delay time is modulated.

Depth

How many milliseconds the delay time is modulated by.

Mix

The wet/dry balance. A flange works because of the mix of wet and dry signal, so the effect will be strongest when this is set to 50%. If this is set to 100%, as all you will hear is the pitch modulating signal, the label changes to "Vibrato."

Gain

Allows you to change the level of the signal.

Pitch Shifter

This allows you to change the pitch of a signal without changing the speed. Parameters include:

Pitch

The amount of pitch shift.

FFT Size

This plugin uses a process called Fast Fourier Transform (FFT), which sub-divides the signal into time (sample) bands. The higher this value, the better your frequency resolution (however, this is at the expense of processing power). If this is set too low, you're likely to hear a "washy" sound (similar to that you get with a low bitrate mp3 file).

Overlap

The longer the FFT size, the better the frequency resolution. However, the length of a band determines the shortest event that can be resolved. Therefore, to increase the time resolution, the FFT time bands are overlapped.

Max Channel

This is related to the number of channels and determines the amount of memory allocated to the plugin. As a general rule, leave this set to 0.

As you can see, to understand all of the parameters of this pitch shift plugin, you'll need quite an in-depth knowledge of digital signal processing. It is also very processor hungry, so use this effect with caution!

171

Transceiver

This allows you to send audio from one Event into another (we used this in Chapter 19). The plugin can be set to transmit or listen to one of 32 channels.

For example, you could create a multi-track music Event, and route each Instrument to a different Transceiver Channel. Next, create separate Events for each stem (each with a Transceiver set to a different channel), and assign them to different FMOD Event Emitters placed around the Unity scene. You would then have diegetic music, which can be mixed by moving around the game world.

A couple of things to note:

- You will need an FMOD Event Emitter in Unity to trigger the Music Event.
- The Advanced Controls > Preload Sample Data tick box must be checked on this Event Emitter in the Unity Inspector.

(You could get a similar result by using separate Events for each stem. However, in order to ensure that the stems are synchronized, they would all have to be triggered at exactly the same time.)

Tremolo

This effect uses amplitude modulation, controlled by an LFO (Low Frequency Oscillator). Parameters include:

Frequency
The frequency of the LFO.

Depth
The amount of amplitude modulation.

Shape
Morphs the LFO waveform between triangle (0) and sine (1).

Skew
Morphs the LFO waveform with a sawtooth – 100% results in a saw-tooth that fades out; +100% results in a sawtooth that fades in.

Square
Controls the "flatness" of the LFO waveform. When set to 100%, this results in a square wave. Lower values give slower on/off transitions.

Duty
Sets the "on" duration of the square wave. Low values will give short spikes.

Spread
Offsets the LFO phase between the left and right channels, giving a spatial widening effect.

Phase
Sets the initial phase of the LFO. For example, when set to 0% on a triangle wave with 100% depth, the output amplitude will start silent. When set to 50%, the output amplitude will start at peak.

Plugin Presets

Right-clicking on a plugin and choosing "Convert To Preset" will convert a plugin to a Preset.

This will be indicated underneath with a circle and arrow icon. Click on this to open the Preset Browser, where you can rename and organize your presets into folders.

FMOD Plugin Presets are not the same as you've seen on a DAW – instances of the Preset are linked.

- Changes made to a preset are applied to all instances of the Preset – even if they are on other Events.
- If automation is used on a Preset parameter, the automation is also applied to all instances of the Preset. This means that you can use one Unity parameter to control multiple FMOD Events.

If you wish to make a Preset plugin independent, right-click and choose "Detach From Preset."

As you've just seen, FMOD plugins provide you with many tools to alter and shape your sounds. In Chapter 19, we'll take a more in-depth look at the Transceiver plugin, and see how we can use it to add an ambience to your Scene.

Ambience and Transceivers

Learning Outcomes

By the end of this chapter, you will be able to:

- Record ambiences for use in a game.
- Configure 3D ambiences in FMOD and Unity.
- Configure Transceivers in FMOD and Unity.

One thing that we've not really looked at yet is adding ambience to your game. This can act as the foundation to the mix – once you get the ambience right, everything else can begin to sit together.

Recording Ambience

Though you can get ambiences and room-tones from an effects library, there's no substitution to recording your own. I carry a portable recorder with me wherever I go – even on holiday – you never know when you might stumble across a great sound effect or ambience. Keeping one of these to hand has enabled me to capture recordings of jungles, oceans, and cityscapes for my library.

If you're planning to record a specific ambience, then what you'll need is:

- Something that can record in stereo
- Batteries (and, ideally, spares)
- Memory cards (or drive space)
- Headphones
- A camera tripod or portable mic stand

Next, you need to decide on your location. Obviously, this depends upon the type of sound you're trying to capture, but for this example, we'll look at recording a quiet outdoor ambience.

Before you head off, look up the location online.

- Are there any near roads or train tracks?
- Perhaps it's near a school? This would be much noisier during the week than at the weekend – but then you might have to deal with the sounds of people mowing their lawns.
- Traffic noise is a significant issue as it tends to be continuous, and is difficult to remove from your recordings.

The time of your recording also makes a big difference – I usually get the best results early in the morning. Once you get to the location, you'll need to critically evaluate the ambience. Don't just concentrate on what you're trying to capture – focus on what you don't want…

The human brain constantly filters out sound that it doesn't think is important, so what may sound on first impression as a fantastic location may have unwanted elements that are included in your recordings – for example, for an indoor ambience, you might not notice the sound of the refrigerator. This "filtering" is called the Cocktail Party Effect. One of the ways that the brain archives this is by analyzing the differences between sounds at each ear (binaural hearing).

If your recording equipment has a mono monitoring capability, you can use this to your advantage. Rather than listen to the actual sound, listen to what is picked up by the microphone. This will make the unwanted noises much more prominent. If you can get rid of the noise source, fantastic. Otherwise, you're going to have to find somewhere else, or come back another time (in the case of a fridge freezer, it's a common trick to turn it off, and leave your car keys inside. That way you'll remember to turn it back on again at the end of the session).

Once you're happy with your location, it's time to set up for recording.

Although you can hold the recorder in your hand, you'll get much better results by placing it on a stand. Even if you try to hold the recorder absolutely motionless, it will pick up handling noise and clothing movement. Most portable recorders have a camera tripod fitting, so this is perfect for the job. Don't forget your wind jammer – even on a calm day, wind turbulence will be picked up.

Now set your levels. If your recorder has an automatic level setting, turn this off. This can lead to "pumping" of the noise, as the signal levels change. In a quiet location, you're likely to have to set your gain quite high.

Hit record, and say your "Slate" – simply say what's being recorded. In the first recording of the day, state:

- The date and time of the recording.
- The location of the recording.
- The equipment used.
- What you're recording.

For later recordings, you can leave out some of these details, but you should make sure that you at least state what's being recorded – this will make life much easier when you get round to organizing your media. For example, a Slate might say:

- 29/01/2018
- Ruislip Rugby Club Grounds
- Zoom h4n built-in stereo microphone
- General Ambience 1

Now stand back, and quietly wait. Get a decent distance away – it's really difficult to stay totally still for the entire recording. I tend to record for at least 3 to 4 minutes, so that I've got enough to work with at the edit.

If anything of interest happened during this time (perhaps there was an airplane fly-past?) then don't stop the recording – it might come in use at later date. Before you press Stop, record a "tail slate" to remind you what you've captured. Now repeat the process several times.

One trick is to turn the recorder around 180 degrees, and capture the ambience of the opposite direction. This can be useful if you later need to fake a surround ambience. It's also worth moving around – if you're recording in the middle of a field, then take the time to make some recordings closer to the edges.

Once you've made your recordings, create your loops (see Chapter 5). Make sure you edit out any distinct sounds as these can make loop repetition far more obvious.

Putting Your Ambience in the Game

When you create your FMOD Ambience Event, it's tempting to set this to 2D, as it will be audible throughout the level, and there is no obvious sound source. However, this can be problematic – for example, if the wind is going from right to left, when the player turns around, the wind should now be heard blowing from left to right. We'll be better off setting it to 3D, but there are a few additional steps we'll need to take.

Create a 3D FMOD Event called "AmbienceEmitter," and place it in the Props folder. Use your ambience loop file (or choose one from the book resources) in a Single Instrument, and set up a Loop Region over the Instrument. Route the AmbienceEmitter Event through the Ambience Group (which you set up earlier to feed the SFX_ HitFiltered Group).

Select the Master Fader, and in the Deck, set the Distance Attenuation to "Off" – this way the level of the ambience will not fade as you move away. You'll also need to set the Max Distance so that it covers the entire game area.

Save the FMOD Project, build your Banks and head to Unity. Right-click in the Hierarchy and choose Create Empty. Name this empty GameObject "Ambience." You can now add an FMOD Studio Event Emitter. Set this to play the AmbienceEmitter Event on Object Start, and try it out in the game.

It will sound odd – especially on headphones. Although the sound does not get louder or quieter as you move around the level, the panning effect is quite extreme. This is where you need to change the Sound Size (see Chapter 6), so return to FMOD, change the Envelopment to User, and adjust the Spatializer Sound Size (this is a great chance to use Live Update for real-time mixing).

Transceiver

If you wish to place smaller ambiences around the scene, then this is quite easy – just change the setting of the Spatializer so that it is only audible for a set distance around the emitter. However, what if the area you wish to cover is not a sphere? You could use multiple Events for this, but there's a really useful plugin we can use so that the sound of one Event can be emitted from multiple locations around the game. It's called the Transceiver.

- Lower the Max Distance and set the Distance Attenuation of the AmbienceEmitter Event back to Linear Squared.
- Right-click in the space above the Master Volume pot and choose Add Effect > Transceiver.
- Set the plugin to "Transmit." It can now transmit audio on any one of 32 channels (numbered 0 to 31). We've not used any other Transceivers yet, so we can leave this on channel 0.
- By default the plugin is set to lower the transmitted signal level by 9dB, so it's worth setting this to 0.

All we now require is the receiver, so:

- Create another 3D FMOD Event called AmbienceReceiver in the Props folder.
- Assign it to the Battleground Bank.
- Route it through the Ambience Group.
- Add a Loop Region in the Logic Track (it doesn't matter how long it's set for, but because there's not going to be any Instruments used, without it, the Event will stop as soon as it is triggered).
- Select the Master Track.
- Set the Spatializer Envelopment to User, and turn up the Sound Size.
- And add a Transceiver plugin before the Volume pot.
- Set the Transceiver plugin Level to 0dB (the plugin will default to Receive on channel 0).
- Save the FMOD project and Build the Banks.
- Return to Unity.

In Unity, create another empty GameObject named "Ambience2," and add an FMOD Studio Event Emitter component. Set this to play the AmbienceReceiver Event on Object Start.

You can then use these overlapping Events to cover an area of the scene.

So what was the advantage of using the Transceiver? Well, as you only have one Instrument playing the sound effect, it is possible to make alterations to the sound without having to copy the same changes to every Event every time.

Our scene now is getting quite busy, with multiple sonic elements, so we're going to have to start mixing the game. Going back and forwards between Unity and FMOD whenever we want to make changes to the mix is awkward, so in Chapter 20, we'll take a look at how we can make this much easier, by connecting Unity and FMOD live together.

20 Real-Time Mixing

180

Learning Outcomes

By the end of this chapter, you will be able to:

- Connect FMOD and Unity projects together to allow for live, real-time mixing.
- Meter the Loudness levels of your game mix.

At the moment, to alter your mix you have to make your changes in FMOD, Save and Build the Banks, before re-launching the game in Unity. This is awkward and inconvenient – but, fortunately, there's a solution to this – Live Update.

Launch Unity, and open the Battleground Scene.

At the moment, as soon as you leave Unity the game pauses – which isn't ideal for mixing. Go to Edit > Project Settings > Player, and in the Inspector, under the Resolution and Presentation section you'll see an option for "Run In Background." Check this, and the game will continue when you change between applications.

Staying in Unity, from the FMOD menu at the top, select FMOD > Edit Settings. In the Inspector you'll see a section labeled "Play In Editor Settings." Check that the Live Update setting is enabled, and take a note of the port number underneath.

Now start the game playing (the game has to be running to connect), and change over to FMOD. At the bottom right is a button labeled "Live Update Off." Click here to open the Connect To Game menu.

If you're running FMOD and Unity on the same computer, you can leave the IP Address set to Local Host. Click on "Connect," and you're up and running.

You can even connect FMOD and Unity running on separate computers. This is great for playtesting and mixing – one person can play the game, while the other mixes the sound.

Enter the IP address of the computer running Unity in the Connect To Game IP box, followed by a colon, then the port number you noted down in Unity.

Finding your IP Address

Mac OSX:
Go to System Preferences > Network, and you'll see the IP address in the Status section.

Windows 10:
Right-click the Start button, select Task Manager and then select the Performance Tab. Choose between Wi-Fi or Ethernet to see the IPv4 and IPv6 addresses.

181

Now you're able to mix in real time, there are a few things to be aware of:

- Your changes won't be "printed" until you save and Build your FMOD Banks.
- Selecting multiple faders will move them to the same level. To maintain their relative levels, hold the shift key before moving any faders.
- By default, FMOD will reconnect to Unity whenever the game is played. Click on the Live Update button (it will change to say "Live Update Reconnecting") to disable this.

Loudness

If you've been following recent developments in the audio industry, you'll have seen that loudness has become extremely important.

So what is loudness? Simply put, it's to do with how loud things sound – or, more importantly, it's used so that we can make things sound as loud as each other.

Human hearing isn't linear – we hear different frequencies with different sensitivities. As well as this, the level of a signal over time also affects our perception – a game that spends most of its time with high audio levels will be perceived as much louder than a game that has just occasional loud sections.

This led to people trying to make their music sound louder than everyone else – and when loudness is your aim, you sacrifice other elements of audio quality.

In response to this (and partially to prevent adverts from sounding much louder than TV programs), film, TV, radio, and now music and video streaming platforms have implemented Loudness Normalization standards (the best known of which is called EBU R128, developed by TC Electronics). The perceived level of the program material is measured and adjusted, so that everything sounds equally loud.

There isn't yet a definitive Loudness standard for game audio, but Sony specify that games for the Playstation 4 should be at -24LUFS. Let's see how to use the FMOD Loudness meter to achieve this.

Go to the FMOD Mixer, and select the Master Bus on the right-hand side to see its settings in the Deck.

Click and hold on the plus button to the right of the volume pot, and select Insert Effect > FMOD Loudness Meter.

We'll need to start with a few definitions:

- 0dBFS (decibels, Full Scale) is the maximum signal level in a digital system.
- 0LUFS (Loudness Units, Full Scale) would correspond to a white noise signal metering 0dBFS.
- A change of 1 Loudness Unit is equal to a 1dB change.

Sony specify that the integrated loudness value of a game should measure -24LUFS, i.e. 24dB quieter than white noise @0dBFS.

To ensure this, we set the Target value to -24LUFS. This means that the zero value on the meter corresponds to -24LUFS.

The orange and blue meter on the left shows the Momentary Loudness value, which gives the average Loudness value over each 400 milliseconds of program material (though this can be changed to Short Term [3 seconds average] by right-clicking on the meter).

Right-clicking also allows you to switch the meter between EBU+9 and EBU+18. You can think of this as the zoom setting of the meter – it doesn't change the value shown, only the range that is displayed.

The Integrated Loudness (Loudness of the entire program material) is shown at the top right, and also as a small triangle to the right of the orange and blue meter. This is the important value!

By default, this Integrated Loudness display is set to Absolute Scale. This means that when the mix is at the right level, it reads -24LUFS. If you right-click here, you can change it to Relative Scale. This will show the LU value relative to the target – for example, if you're 6dB louder than the target (-18LUFS), it will display +6LU. Once you've decided upon your target loudness, this can be a much better way to display your readings.

The Range value shows the differences between the loudest and quietest sections. It's also visible as a gray line to the side of the orange and blue meter.

Max Peak shows the true peak sample level. Due to interpolation in digital audio, it is possible to have a peak higher than 0dBFS. This is not allowed by most audio standards, so Loudness meters use over-sampling to detect these. Any value higher than 0dBTP (decibel True Peak) indicates inter-sample clipping.

183

Inter-Sample Clipping

Digital audio is time-discrete. This means that we do not have amplitude values for all points in time, but only at regular intervals. These intervals are determined by the sample rate (e.g. a sample rate of 48kHz means that we have 48,000 amplitude data points per second).

However, when the audio waveform is reconstructed, we don't draw a straight line between each point. Instead, we interpolate a smooth curve. Think of this as akin to a join-the-dots puzzle – if you were to complete the picture by inter-connecting all the dots with a ruler, you'd end up with a jagged, ugly end result. Even a small child learns that interpolating smooth lines gives a much better reproduction of the original image.

This can lead to inter-sample clipping, where the interpolated curve goes above 0dBFS.

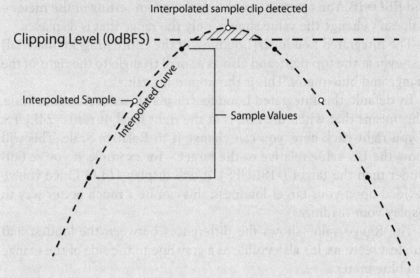

Figure 20.1 Inter-sample clipping

True peak meters use oversampling to interpolate data points between the samples, which can then be checked for any clipping.

We're very likely to see Loudness standards introduced in game audio, so it's definitely worth using Loudness metering in your mixes.

Setting up a basic mix is quite straightforward, but we're going to need our mixes to change as the player moves through and interacts with the game world.

In Chapter 21, we'll see how to configure this with the use of Mix Snapshots.

Mix Snapshots

Learning Outcomes

By the end of this chapter, you will be able to:

- Create and configure FMOD Mixer Snapshots.
- Create Reverb zones in the Unity Scene.
- Trigger FMOD Mixer Snapshots inside Events.

Although the avatar's footsteps are set to change when the player enters the Fort, it will add a touch of realism to add reverb. This could be done via parameter control, but a much simpler solution is to use Mix Snapshots and Reverb Zones.

This chapter will take you though setting up Mix Snapshots in FMOD, setting up Colliders in Unity to trigger the snapshots, and using Layers in Unity to prevent the Rockets from colliding with the Reverb zone.

We'll also see how to configure an FMOD Mix Snapshot to briefly apply a low pass filter when the player is shot by a Drone.

Let's start by configuring our snapshot in FMOD. Snapshots save and recall your mix settings, and they can be triggered by Unity just like Events. A game mixer will create snapshots for the different states in the game, for example:

- Different game environments.
- Different health conditions.
- To change the "emotion" of the mix – for example, lowering the music and bringing up the foley during a stealth section can increase the player's feeling of suspense and tension.
- When the game is paused (e.g. if the player switches to the settings menu or a map).
- When a battle starts in the game.

These snapshots act as the basis for the mix, which can then be dynamically tweaked and adjusted using parameter controls and side chains etc.

Open the Player Mixer view in the FMOD Mixer. If you've been experimenting with auxiliary routing, make sure that the Reverb Send property is turned all the way down on the Player Group (this was set up in Chapter 17).

On the left side of the window there are Tabs available for Routing, VCAs, and Snapshots. Change to the Snapshots Tab, and at the bottom of the page, click on "New Snapshot" (this option is also available by right-clicking in the gray-space of the Tab).

You are offered two options: New Overriding Snapshot and New Blending Snapshot. The only difference between these two is in how they affect the Volume properties.

An Overriding Snapshot will change all "scoped" properties (we'll come to this in a moment) to the snapshot values.

A Blending Snapshot will do exactly the same, except for the Volume properties, which will blend additively (for example, if the Snapshot Volume property is set to +6dB and the current Volume is set to -4dB, when a Blended Snapshot is recalled, the Volume will be set to +2dB).

We're not going to be including the Volume property in our snapshot, so it doesn't matter which of these you select on this occasion. Name the Snapshot "InFort."

You'll be able to see that the Reverb Send pot and the fader caps for the Groups, Returns, and VCAs are now shown as dotted lines. This is because these properties are not currently "scoped in" – i.e. they will not be affected by the Snapshot (the Event faders are still shown with yellow caps because they cannot be included in Snapshots).

Click on the Reverb Send pot (or right-click and choose "Scope In") and it will fill in to indicate that it is scoped. Turn the pot up, and your Snapshot will now add reverb to the Player Group when it is triggered.

If you need to leave the snapshot and go back to the default mix, click on where it says "Mixing Desk" (just under the Transport controls).

Save and Build your project, and head over to Unity.

Select the Fort in the Hierarchy, move your cursor over the Scene, and press [F] to focus on the Fort.

We're going to be creating a Reverb Zone – an area of the game world where reverb is added to the mix. We could do this by adding a Collider directly to the Fort, but we'll see later that there will be issues if we do it this way (we'll need to ensure that the Fort and Reverb Zones are in separate Layers).

- With the Fort still focused in the Scene, go to GameObject > Create Empty. Name this "FortReverbZone."
- In the Hierarchy, drag the FortReverbZone onto the Fort, making it a child (this will ensure that the Reverb Zone moves with the Fort if you ever wish to re-arrange your Scene).
- Set the FortReverbZone Transform values to 0,0,0 to ensure it is exactly on top of the Fort.

187

There are several options for our collider shape, but none of Unity's primitive colliders are ideal here. You could approximately cover the Fort with multiple box colliders, but as the model is relatively simple, we can get away with using a Mesh Collider. Go to Add Component > Physics > Mesh Collider (or type "mesh" in the search menu).

Click on the icon of the dot in a circle to the right of the Mesh box, and you will see a list of all meshes included in your Unity project. Look for "Fort," and double-click to set this.

As it stands, this won't work – the Collider will only detect if you are inside the walls, and nothing will be able to pass though it to enter the Fort. To fix this:

- Set the Collider to "Convex" (this gives us a simplified mesh that covers the entire Fort).
- Set the Collider to "Is Trigger" (objects will now be able to pass through the colldier).

Now go back to Add Component and add an FMOD Emitter (it's under the FMOD Studio sub-section). Set:

- The Play Event to "Trigger Enter."
- The Stop Event to "Trigger Exit."
- The Collision Tag to "Player."
- The Event to Snapshots > InFort (click on the magnifying glass icon).

Trigger Enter vs Collision Enter

Trigger Enter (what is referred to in Unity as OnTrigger Enter) is similar to Collision Enter (OnCollisionEnter), but instead of being called when two Colliders hit, it is called when a RigidBody enters a Collider set to "Is Trigger."

Unlike a Collision Enter, it is possible for a Rigidbody to pass through a Trigger. In addition, Trigger Enter on an FMOD Studio Event Emitter offers the option to set a Collision Tag, which determines which GameObjects it will be affected by.

Try out the Reverb Zone in the game – when you walk into the Fort, you should hear reverb on all the Player sounds.

Try firing a Rocket from the Fort – you might notice that there is a problem…

So what happened? The Collider works to trigger the snapshot; however, it's also causing the Rocket to explode.

To avoid this, we're going to put the Rockets and the Reverb Zones into separate Layers. We can then configure Unity so that these Layers will not interact.

Go to Edit > Project Settings > Tags and Layer, and in the Inspector, click on the disclosure arrow to view the Layers. You can see that there are eight "Built-in" layers (numbered 0 to 7), and spaces to add more of your own. Type "RocketLayer" and "NotRocketLayer" into User Layers 8 and 9.

Select the FortReverbZone in the Hierarchy, and at the top right of the Inspector, change its Layer from Default to NotRocketLayer.

Navigate to and select the Rocket Prefab (it should be under the "MyPrefabs" folder), and change it to the RocketLayer.

Finally, we'll need to configure how Layers interact with each other. Go to Edit > Project Settings > Physics.

In the Inspector, you'll see a matrix of check boxes for the Layer collisions. Find the vertical column for the NotRocketLayer, follow it down to the RocketLayer row, and de-select the check box. These two layers will no longer interact.

Test this out in the game. You should still hear reverb when you enter the Fort, but you will now be able to fire Rockets through the collider.

You can also set the RocketLauncher to the NotRocketLayer. This will prevent the issue we had earlier (where the Rockets were colliding with the launcher), so you can re-position the ShootFromHere closer to the mouth.

That's working nicely, but it's not the only way that snapshots can be used – they can also be included inside FMOD Events. We'll demonstrate this by adding a low-pass filter to the mix that is triggered whenever the player is hit by a laser.

189

- Head back to the FMOD Mixer window, and add a Multiband EQ to the SFX_HitFiltered Group.
- Push the low pass filter in band A up to 22kHz, so that it doesn't yet have any effect.
- Create a new Snapshot (overriding or blending), and name this "HitFilter."
- Click on the Frequency property to scope it in to the snapshot. Set this to around 1kHz (we can fine-tune this later).
- Change to the Event Editor window, and create a new 2D Event in the Player > PlayerImpacts folder called "LaserHit." Assign this to the PlayerBank.
- Right-click in the Audio Track, and Add Snapshot Instrument > HitFilter. Drag out the Trigger Region so that it covers from 0:00 to around 3 seconds, before saving and Building the Banks.
- In the Mixer window Routing Tab, drag the LaserHit Event into the SFX_NoHitFilter Group.

We could trigger this Event in several ways, but let's keep it simple, and use an FMOD Studio Event Emitter.

- Go to Edit > Project Settings > Tags and Layers, and add a tag called "LaserTag."
- Select the LaserBolt prefab (under MyPrefabs), and change its Tag to LaserTag.
- Select the FPSController in the Hierarchy, and add an FMOD Studio Event Emitter.
- Set the Play Event to "Trigger Enter."
- Leave the Stop Event as "None."
- Set the Collision Tag to "LaserTag."
- Set the Event to event:/Player/PlayerImpacts/LaserHit.

Play the game, and stand in front of a SentryDrone. When it shoots you, you will hear a low pass filter on the mix, that stops abruptly after around 3 seconds.

Let's make the snapshot gradually fade out. There are again several techniques we could use, for example:

Technique 1

- In the LaserHit Event, move the mouse to the top right of the Snapshot Instrument Trigger Region. You will see it change to a fade tool, which you can use to set a fade-out.

Technique 2

- Select the Snapshot Instrument, and right-click on the Intensity pot, and Add Modulation > AHDSR. This gives you an intensity envelope (we looked at these in detail in Chapter 16b).
- Set the Attack to 0 seconds, and adjust the Release to whatever you find works in the game.

It's entirely up to you which technique you use, but the FMOD manual recommends the use of AHDSR modulators to ensure smooth snapshot transitions.

You can take your LaserHit Event even further – perhaps you could add an Audio Track with a vocalization when the player is hit? (as we placed the LaserHit Event in the SFX_NoHitFilter Group, it will not be affected by the Snapshot). You'll find several VO_Avatar_Hit audio files for this in the book resources.

In Chapter 22, we'll be improving the realism by adding ambience to the Scene.

191

22

The Health System

Learning Outcomes

By the end of this chapter, you will be able to:

- Add a health system to the game.

Over the next two chapters, we'll be creating a health system for our game. Let's define how we want it to behave:

- We'll assign a number of hit points to the player.
- Every time the player is hit, they lose 1 hit point.
- There will be a number of health pickups scattered around the game world.
- When the player walks into one of these, the health pickup is destroyed, a sound effect is triggered, and the player regains a hit point.
- When the player is down to a single hit point, a fast beating heartbeat will be heard.
- When all the hit points are lost, Unity will load a "Game Over" Scene, with an option to restart the game.

Before we jump straight into scripting, let's get everything else ready.

Health Pickup

We'll need at least one health pickup to test our script, so import the HealthPickup.fbx model to the MyModels folder, and extract its textures and Materials (see Chapter 13 if you've forgotten how to do this).

Drag this into the Scene, somewhere close to the FPSCharacter.

Next, go to the Edit > Tags & Layers, create a new Tag called "FirstAid," then re-select the HealthPickup in the Scene, and apply the new tag.

We'll need to be able to detect collisions with the HealthPickup, so go to Add Component > Physics > Sphere Collider. Set this to "Is Trigger."

To make the pickup stand out, find the SpinScript (it should be in the MyScripts folder), and drag it onto the HealthPickup to make it spin around.

Finally, navigate to the Project Tab MyPrefabs folder, and drag the HealthPickup over from the Hierarchy to make it a prefab (we'll need several of these in the game).

Game Over Scene

Navigate to the MyScenes folder, and right-click in the Project Tab gray-space to Create > Scene. Name the Scene "GameOver," but don't open it – we just need it as a placeholder for now (we'll finish setting it up in Chapter 23).

However, before this can be used, we need to tell Unity that the GameOver Scene is part of the same game.

Open the MyScenes folder in the Project Tab, then go to File > Build Settings. Drag the Battleground and GameOver Scenes into the "Scenes In Build" section of the Build Settings window. (Note that the order of these is important – the Scene at the top will be the first one that is loaded when you launch the game, so make sure that this is the Battlegrounds Scene. You can always click and drag to re-arrange the order.) You don't have to do anything else in here, so you can close down the window.

FMOD Health Events

In the Player > Vocals folder, create a new 2D Event called "Heartbeat" (I know, it's not really a vocalization, but it's the closest category we have), and assign it to the PlayerBank.

Let's take this opportunity to see another way to bring sounds into FMOD: you can drag from the browser, straight onto the Audio Track. This will automatically create a Single Instrument, and import the audio file to the FMOD project (if you drag multiple audio files at once, it will create a Multi Instrument).

Drag the Instrument to the start of the Timeline, and right-click to create a new loop region that matches the Trigger Region length.

Create a new 2D Event in the Props folder, and name this "HealthPickup." Use the same technique we just saw to bring in the SFX_Pickup_Health01 file to a Single Instrument, and assign this to the PlayerBank (we don't need a loop region for this Event).

Neither of these sound effects should be affected by the HitFilter, so open the Mixer Window, and place them in the SFX_NoFilter Group.

That's everything we need in FMOD for now, so, as ever, Save and Build your Banks, and return to Unity.

Open the MyScripts folder, and right click to Create > C# Script. Name this "HealthSystem," and drag it into the FPSController (in the Hierarchy), before double-clicking to open it in your IDE.

Copy over the following code:

```
 1 using System.Collections;
 2 using System.Collections.Generic;
 3 using UnityEngine;
 4 using UnityEngine.SceneManagement;
 5
 6 public class HealthSystem : MonoBehaviour
 7 {
 8     public int hitPoints = 3;
 9     private int healthCheck;
10     private FMOD.Studio.EventInstance heartbeat;
11
12     // Use this for initialization
13     void Start()
14     {
15         healthCheck = hitPoints;
16         //print ("hitPoints = " + hitPoints);
17     }
18
19     void OnTriggerEnter(Collider other)
20     {
21         if (other.gameObject.tag == "LaserTag")
22         { // determines tag of objects that will cause score to decrease
23             hitPoints = hitPoints - 1; // decrease hitPoints by 1
24                             //print ("hitPoints = " + hitPoints);
25             if (hitPoints == 1)
26             {
27                 heartbeat = FMODUnity.RuntimeManager.CreateInstance("event:/Player/Vocals/Heartbeat");
28                 heartbeat.start();
29                 //print ("heartbeat has started");
30             }
31             if (hitPoints == 0)
32             {
33                 heartbeat.stop(FMOD.Studio.STOP_MODE.ALLOWFADEOUT);
34                 FMOD.Studio.Bus musicBus = FMODUnity.RuntimeManager.GetBus("bus:/Music");
35                 musicBus.stopAllEvents(FMOD.Studio.STOP_MODE.IMMEDIATE);
36                 SceneManager.LoadScene("GameOver");
37             }
38         }
39
40         if (other.gameObject.tag == "FirstAid" && hitPoints < healthCheck) // determines tag of objects that will cause score to increase
41         {
42             hitPoints = hitPoints + 1; // increase hitPoints by 1
43             Destroy(other.gameObject); // destroy the Health Pickup
44             FMODUnity.RuntimeManager.PlayOneShot("event:/Props/HealthPickup"); // play the FMOD HealthPickup Event
45             //print ("hitPoints = " + hitPoints);
46             if (hitPoints == 2)
47             {
48                 heartbeat.stop(FMOD.Studio.STOP_MODE.ALLOWFADEOUT);
49                 //print ("heartbeat has stopped");
50             }
51         }
52     }
53 }
```

Figure 22.1 Health script

HealthSystem Script Walkthrough

At line 4, you can see that we're using a new class called "SceneManagement." This gives us access to Methods that can switch between different Unity Scenes – in this script, we'll need this to open the "GameOver" Scene.

Lines 8 to 10 are where we declare our variables:

- "hitPoints" is the maximum health of the player. It has a default setting of 3, but as it's been set as a Public variable, you can change this from the Unity Inspector.

- "healthCheck" is used to ensure that health pickups only work if the player has lost hit points (see line 40).
- "heartbeat" is used to trigger and stop the FMOD Heartbeat Event.

Start Method

In here, we set the healthCheck variable to be the same as the hitPoints value (in case you've changed this in the Unity Inspector).

OnTriggerEnter Method

This is used for both being hit by a Laser, and when finding a health pickup.

LaserTag

If the GameObject that the script is placed on collides with another gameObject tagged as "LaserTag," then the hitPoints value is lowered by 1 (hitPoints = hitPoints − 1).

Nested under this condition are two other "if" statements:

If the hitPoints value is equal to 1 (the double-equals sign means "is equal to," as opposed to a single-equals sign, which is used to assign a value to a variable), then the FMOD Heartbeat Event is triggered.

If the hitPoints value is equal to zero, then:

- The FMOD Heartbeat Event is stopped.
- A new FMOD Studio bus variable is declared (musicBus), and set to the FMOD mixer Group called "Music."
- All Events going through the FMOD "Music" Group are stopped immediately (without being allowed to fade out).
- The GameOver Scene is loaded.

At line 34, you can see that the FMOD.Studio.Bus musicBus variable is declared within the OnTriggerEnter Method. There's nothing wrong with this – some coders prefer to declare variables as and when they are needed. This also can allow the code to be more efficient (the variable can be forgotten about once the function has completed). However, one disadvantage is that it can then only be used inside the function in which it is declared (for example, I couldn't use the musicBus variable inside the Start Method).

Of course, you won't actually hear the effect of this until we add music to the game (we'll start working on this in Chapter 24), but we set things up ready for this back in Chapter 17.

If we'd set this to affect the Master Bus, (FMODUnity.Runtime Manager.GetBus("bus:/");) then this would stop all FMOD events – which could result in sudden truncations of some Events.

FirstAid
If the GameObject that the script is placed on collides with another gameObject tagged as "FirstAid," then:

- The hitPoints value is increased by 1.
- The other gameObject is destroyed.
- The FMOD HealthPickup Event is triggered.

Nested in this condition is another "if" statement:

- If the hitPoints value is equal to 2, then the FMOD Heartbeat Event is stopped.

You can also see I have left comments in the code. Some of these are to remind me (or to tell other programmers) how the code works, but the "print" commands were from testing out the script. These are helpful to narrow down exactly where any mistakes have happened: for example, how would I know if the reason I can't hear the heartbeat is due to mis-typing a Tag, or because I made a mistake in FMOD? These can at least show me if the game is trying to trigger the Event. If you need to use any of these, delete the double forward-slashes at the start of the line.

While we're talking about comments, let's see how to comment out entire sections, rather than line-by-line.

/* and */ will comment out all code in between. This can be used within a line, or across multiple lines.

This is not a particularly common technique, as it's not as obvious to the eye as double-forward slashes.

If you need to comment in and out large sections, select all of the lines, hold [cmd] (Mac) / [ctrl] (Windows), and type [/] (note that this has to be the forward-slash on the alpha-numeric keypad, not the numeric keypad).

Try this out in the game. Stand near a SentryDrone, and let it shoot you.

- When you're down to 1 health point, you should hear the heartbeat sound effect.
- Run into the HealthPickup, and hopefully the heartbeat will stop.
- Once your hitPoints are down to zero, Unity should load the GameOver Scene (and you'll probably see a number of error messages – don't worry about these, we'll deal with them in Chapter 23).

If it's not working, you'll need to narrow down where you've made a mistake. Common issues to check for include:

- Silly "brain fart" mistakes (e.g. did you remember to apply the HealthSystem script to the FPSController?).
- Spelling mistakes: have you used the exact same spelling for your Tags, variables, and FMOD Events?
- Upper-/lower-case mistakes: Unity is case sensitive (e.g. if you've tagged your HealthPickup as firstAid, and the code is checking for FirstAid, then it won't work).
- Missing brackets: all opened brackets in your code must be closed! Selecting an open curly bracket in your IDE will highlight the corresponding close bracket.
- Missing semicolons: all single-line statements in C# must end in a semicolon. Look for red squiggly underlining in your IDE to indicate these.
- Messed up quotation marks: in a word processor document, there is a difference between open and closed quotation marks ("like either side of this"), unlike a plain text document ("where they will look like this"). These will cause problems if you're copying and pasting code.

Almost all of these are easy mistakes to make, and can often be difficult to pin down. When you combine this with the occasional bug in FMOD or Unity, which could cause the game to break for no apparent reason, it's also important to make regular backups of your work.

In a professional development environment, you'd use a Version Control system such as Git or Perforce. These keep backups of all versions, and document who made which changes. It should then be possible to go back to a version of the game before the error happened (and know who to blame!). They also make it a lot easier for multiple people to work on one project.

As a sound designer, you're unlikely to have access to any critical versions, so any mistakes you make shouldn't affect the game – but this is assuming that everything has been set up correctly. As a rule, check with someone before making any changes!

Our game is relatively small, so you should be able to figure out your own backup solution. Each day, before I continue work on a project, I make a backup zip to my RAID drive. This habit has saved me more than once!

Another thing to watch for is changing software versions. As a rule, never do this while you're working on a project (unless it's absolutely unavoidable). Code is changed, depreciated, and made obsolete with almost every update. Unity will try its best to fix this for you, but there's no guarantee it will work (and that's before you get to re-integrating your Unity and FMOD projects).

So, take a moment to back up your work, and in Chapter 23, we'll finish our GameOver Scene.

Updating a Unity and FMOD Project

There are regular updates to FMOD and Unity. However, you should avoid making any version updates while you're still working on a project – they can break your game!

Sometimes though, this is unavoidable – perhaps you want to go back to a game you were working on at an earlier date? I've experimented with this, and found a method that usually works:

- Back up your FMOD and Unity projects (make a safety copy, in case this procedure doesn't work).
- Open the Unity Scene (you will see an error message that the version number of the file format is not supported).
- Go to FMOD > Migration From Legacy Integration. Press OK at the warning message.
- Go to Assets > Import Package > Custom Package, and install the latest FMOD Studio Unity package.

- You will see a console message telling you that the FMOD path has not been set. Do NOT change it yet.
- Quit Unity.
- Re-open the Unity project/Scene.
- Go to FMOD > Edit Settings.
- This will be set to "Multiple Platform Build." Change it back to "Project."
- Browse to and select your FMOD project in the Unity Inspector.
- Open the FMOD project.
- Save and Build your Banks.
- Return to Unity.

This procedure has been successful for me on several occasions, but, as a rule, stick with the Unity and FMOD versions that you start the project with!

Game Over Man!

Learning Outcomes

By the end of this chapter, you will be able to:

- Create a "GameOver" Scene.
- Use 2D Sprites on a Canvas in a Unity Scene.

Before we start setting up our GameOver Scene, let's create our Events in FMOD. We're going to need:

- Music
- An impact/death sound effect
- Keypress sound effects

Don't worry about the Music Event for now – we'll deal with this in a later chapter.

The death sound effect will be fairly straightforward. Create a 2D Event in the PlayerImpacts folder called "GameOverImpact," and use the VO_Avatar_Death_Fall.wav file in a Single Instrument.

Create a new folder in FMOD called "UI_SFX," and inside here create two 2D Events called "KeyPressCorrect" and "KeyPressError." Use the UI_KeyPress audio files in Single Instruments for each of these.

Don't forget to organize the Mixer routing for the new Events – you should already have a UI group ready for the KeyPress sound effects.

We could place the Events in existing Banks, but when we get to optimizing our game, we might want to unload these Events while we're in the Battleground Scene – and this will be easier to do if they're all in the same place.

- Select all three of your new Events, right-click, and choose Assign to Bank > New Bank.
- Name this "GameOverBank."
- Save your FMOD project and Build the Banks.
- Head over to Unity, and open up the GameOver Scene.

At the moment, there's not a lot here, so let's start by importing in our screen graphic. Create an Assets sub-folder called "MyGraphics," and import the GameOver.png and YouWin.png files from the book resources.

With both the GameObject.png and YouWin.png files selected, take a look at the Texture Type property in the Inspector – it will be set to "Default." As these are going to be used in a 2D User Interface, this will need to be changed to "Sprite," so make the change, and then click on "Apply" at the bottom right.

We don't just drag our graphic onto the scene – all UI elements need to be placed on a "Canvas." We'll then need to place the graphic on the Canvas as part of an "Image." We can add both of these in one go – go to GameObject > UI > Image, and Unity will add a Canvas to the Scene, with an Image as a Child.

With the Image selected in the Hierarchy, drag the GameOver.png sprite into the Source Image box.

As this will be a 2D menu screen, there's not much point working in the Scene view, so change this to Game, so that you are seeing through the Main Camera. You should be able to see the graphic, but it will work better if we stretch this to fill the entire screen.

Select the Image in the Hierarchy, and click on the Anchor Presets box at the top left of the Rect Transform component. Now hold down [option] (Mac)/[alt] (Windows), and you'll see the Anchor Preset options change. Keep the key held down, and choose the bottom right option to make the graphic fill the screen.

By default, Sprites don't use dynamic lights in Unity. This means that our Directional Light isn't doing anything, so we can get rid of it – right-click on it in the Hierarchy, and select "Delete."

Great – we have our GameOver graphics looking good – all we need now is to add our sound effects and script.

If you tested out dying in the Battleground scene earlier, you'll have seen a number of warning messages – for example, it will have complained that there was no FMOD listener when the GameOver scene was launched.

We'll add one of these to the Main Camera, so select this from the Hierarchy, and in the Inspector, go to Add Component > FMOD Studio > FMOD Studio Listener.

I'd like a sound effect to be played when the GameOver scene loads, as well as to start playing music. We could do this by using FMOD Studio Emitters, but, just for a change, let's do this as part of a script. We'll use the same script to restart the game.

Navigate to the MyScripts folder, and right-click in the gray-space to create a new C# script. Name this "GameOver," and double-click to open it in your IDE.

It's copy and paste time again, so grab the code from the GameOverScript1 text file, and paste it in so that it looks like this:

```
1 using System.Collections;
2 using System.Collections.Generic;
3 using UnityEngine;
4 using UnityEngine.SceneManagement;// allows the script to change the Unity scene
5
6 public class GameOver : MonoBehaviour
7 {
8      // Use this for initialization
9      void Start()
10     {
11         FMODUnity.RuntimeManager.PlayOneShot("event:/Player/PlayerImpacts/GameOverImpact"); // play the GameOverImpact Event
12     }
13
14     // Update is called once per frame
15
16     void Update()
17     {
18         if (Input.anyKeyDown)
19         {
20             //print("key pressed");
21             if (Input.GetKeyDown(KeyCode.R) || Input.GetButtonDown("Fire1"))
22             {
23                 //print("button pressed");
24                 FMODUnity.RuntimeManager.PlayOneShot("event:/UI_SFX/KeyPressCorrect"); // play the FMOD KeyPressCorrect Event
25                 FMOD.Studio.Bus musicBus = FMODUnity.RuntimeManager.GetBus("bus:/Music");
26                 musicBus.stopAllEvents(FMOD.Studio.STOP_MODE.ALLOWFADEOUT);
27                 SceneManager.LoadScene("Battleground");
28             }
29             else
30             {
31                 //print("wrong key");
32                 FMODUnity.RuntimeManager.PlayOneShot("event:/UI_SFX/KeyPressError"); // play the FMOD KeyPressError Event
33             }
34         }
35     }
36
37 }
```

Figure 23.1 GameOver script 1

GameOver Script Walkthrough 1

At the top of the script, we can see the usual "using" options – and we've added the SceneManagement class, as we wish to be able to relaunch the Battleground scene.

In the Start Method, we play the GameOverImpact Event.

The Update Method is run once per frame, and it checks to see if any key has been pressed down (this is slightly different to "Input.anyKey," which would continue to trigger while the key is held. Input.anyKeyDown is only called when a key is first pressed).

If the pressed key is [R] or if the left mouse button (Fire1) is pressed, then:

- The KeyPressCorrect Event is played.
- The music is stopped.
- The Battleground scene is reloaded.

|| indicates "OR" in C#. The line symbol is called the Pipe, and hopefully you'll find it using [shift]+[\]. If you can't find it (it can move with different keyboard layouts), then it might be easier to copy and paste it from other documents whenever you need it!

If any other key is pressed, then the KeyPressError Event is played.

Save the script, and drag it onto the Main Camera (we could use any of the GameObjects in the Scene, but we're definitely going to be keeping the Main Camera. It's quite a common practice to use the Main Camera or Directional light to hold scripts, rather than creating dedicated GameObjects).

That's pretty much done. Play the Scene, and try out mouse clicks and different key presses. When you press [R], you should be launched straight into the Battleground.

We're going to need a similar Scene for when the player wins the game (we'll be setting up the win conditions in a later chapter), so while you've got the process fresh in your mind let's take this opportunity to get it ready.

- Navigate to the MyScenes folder, and create a new Scene called "YouWin."

- Double-click on the YouWin Scene to open it (save the previous Scene if prompted).
- Go to File > Build Settings.
- Click on "Add Open Scenes" to add the YouWin Scene, then close the Build Settings window.
- Delete the Directional Light from the Scene.
- Go to GameObject > UI > Image to add a Canvas and an Image to the Scene.
- Drag the YouWin.png sprite into the Source Image in the Inspector.
- [opt] (Mac) / [alt] (Windows) – click on the bottom right Anchor Preset so that the YouWin.png sprite graphic fills the screen.
- Add an FMOD Studio Listener component to the Main Camera.
- Create a new script called "YouWinScript" in your MyScripts folder.
- Attach the YouWinScript to the Main Camera.
- Double-click on the script to open it in your IDE.
- Copy over the code from the YouWinScript text file included in the book resources.
- Save the Script.
- Save the Scene.

```
1 using System.Collections;
2 using System.Collections.Generic;
3 using UnityEngine;
4 using UnityEngine.SceneManagement;// allows the script to change the Unity scene
5
6 public class YouWinScript : MonoBehaviour
7 {
8
9     // Use this for initialization
10    void Start()
11    {
12        FMODUnity.RuntimeManager.PlayOneShot("event:/UI_SFX/YouWin"); // play the GameOverImpact Event
13    }
14
15    // Update is called once per frame
16    void Update()
17    {
18        if (Input.anyKeyDown)
19        {
20            //print("key pressed");
21            if (Input.GetKeyDown(KeyCode.R) || Input.GetButtonDown("Fire1"))
22            {
23                //print("button pressed");
24                FMODUnity.RuntimeManager.PlayOneShot("event:/UI_SFX/KeyPressCorrect"); // play the FMOD KeyPressCorrect Event
25                FMOD.Studio.Bus musicBus = FMODUnity.RuntimeManager.GetBus("bus:/Music");
26                musicBus.stopAllEvents(FMOD.Studio.STOP_MODE.ALLOWFADEOUT);
27                SceneManager.LoadScene("Battleground");
28            }
29            else
30            {
31                //print("wrong key");
32                FMODUnity.RuntimeManager.PlayOneShot("event:/UI_SFX/KeyPressError"); // play the FMOD KeyPressError Event
33            }
34        }
35    }
36 }
```

Figure 23.2 YouWin script

Script Explanation – YouWinScript

Not much to explain here – it's almost exactly the same as the GameOver script. The only differences are in the Start Method – we're going to play a "YouWin" Event.

If you've looked through the script, you'll already have figured out that we need to create a "YouWin" Event. This will be a 2D Event, and should be placed in the FMOD UI_SFX folder (use the SFX_UI_YouWin_Fanfare.wav file in a Single Instrument).

We could create a dedicated Bank for this, but we can get away with placing it in the GameOver Bank. Don't forget to route the YouWin Event through the UI_Vox Mixer Group!

Save the FMOD Project, Build your Banks, and the YouWin Scene is done for now (we'll be coming back to it in Chapter 26).

In the next few chapters, we'll be creating our game music. There's quite a lot to get through, so it might be worth taking a quick break before jumping back in. See you over the page!

Game Music Part 1

Variation

Learning Outcomes

By the end of this chapter, you will be able to:

- Use Multi Instruments to randomize music.
- Use Trigger Behaviors to determine if Instruments are triggered.
- Use nested Loop Regions and Loop Probabilities to randomize music arrangement.
- Use Time.deltaTime to control FMOD parameters.

Music adds an entire dimension of immersion – just hearing the score to *Tetris*, *Mario*, or *The Last of Us* can put the player back in the game world.

However, game music has different requirements to film score. A player could breeze through a level in minutes, or get stuck in the same area for hours. A simple looping score would quickly become obtrusive and annoying – as anyone who's ever worked in retail over Christmas could tell you!

A triple-A title can easily have 30 to 40 hours of game play – and that won't even scratch the surface of an MMO such as *World of*

Warcraft. It's unreasonable to expect 40+ hours of original music to be composed and arranged for a game.

But as we said earlier – repetition breaks immersion. This can be avoided using:

- Compositional techniques: avoid using any obvious melodies or phrases in loops, as these make repetition far more noticeable. This can be accomplished using generative composition techniques such as "random walk," or with slow atmospheric pads.
- Long Loops: the longer the loop, the less likely you'll notice any repetition.
- Randomization techniques: random musical variations are applied to the music playback.

The focus of this book is on audio implementation, so we'll be concentrating on randomization methods. We'll be demonstrating these techniques using the GameOver Scene (as this doesn't require any parameter control), but they can – and should – be also applied to the Battleground music.

Create a new Event folder in FMOD called "Music," and create a new 2D Event in here called "GameOverMusic" (we'll be calling this Event from our code, so be very careful to ensure that the folder and Event names match exactly). Before you forget, route the GameOverMusic Event to the Music group in the FMOD Mixer window.

To save time, import all of the music files to your FMOD project – and while you're here, it will be worth organizing them into a subfolder. Select all of the music files in the Audio Bin, right-click, choose the "Move Into New Folder" option, then name the folder "MusicFiles."

As we're going to be working on music with a known bpm, change the Timeline from "Time" to "Beats" just above the time counter to the right of the Transport controls.

As with pretty much every DAW, this defaults to 120bpm, so we'll have to change it to match the tempo of our GameOver music loops. We do this in the Logic Track (the black area just above the Audio Tracks where we've placed our loop regions).

Right-click (above the Track, not the Track header), and you have several options available. You need "Add Tempo Marker," which creates a 120bpm, 4/4 marker. Drag this to the start of the Timeline, double-click on the marker and change it to 110bpm.

Now your editing will snap to a grid, which makes working on music much easier.

Multi Instrument Randomization

- Name the Audio Track "DrumLoop," then drag over both the GameOverMusicDrumLoop01 and GameOverMusicDrum Loop02 wav files together from the Audio Bin, to create a Multi Instrument.
- Drag this over to the start of the Timeline, and right-click on the Multi Instrument to create a New Loop Region.
- In the Deck, you'll be able to see the Multi Instrument Playlist. Right-click on the GameOverMusicDrumLoop01 Instrument in the Playlist to set the Play Percentage. Double-click, and try out a value of around 75%. Now when you loop the Event, the GameOverMusicDrumLoop02 will only play one time in four.

This technique works best when you layer multiple Audio Tracks – which is why it is sometimes referred to as Vertical randomization.

- Right-click on the Audio Track and Add New Audio Track.
- Name this "BassLoop," and repeat the same process we used for the DrumLoop using the GameOverMusicBassline files.

We now have four combinations of loop files. When you combine this with the fact that the sequence of the combinations is random, repetition will now be far less noticeable. The more layers we add, the more combinations we get – adding another Multi Instrument with just two audio files will double the possibilities.

We can use Single and Multi Instruments to add embellishments – an occasional flourish or effect added to the playback.

You might think that this could be achieved by using a Multi Instrument containing just one Instrument in its Playlist, and using the Play Percentage. However, this won't actually work – if a Multi Instrument contains just one item in the Playlist, it will trigger every time, regardless of the Play Percentage setting. Instead, we're going to use Trigger Behaviors.

Create another Audio Track, give it a name (I'm calling mine "Flourish"), and drag over the GameOverMusicFlourish01.wav file from the Audio Bin to create a new Single Instrument.

With the new Instrument selected, click on the Trigger Behavior disclosure arrow at the left-hand side of the Deck. We'll be looking at most of these controls in Chapter 25, but for now click on the picture of the die on the left to activate the Probability setting. Set this to around 50% (so we don't have to wait too long when testing the Event) and the Instrument now has a 50:50 chance of playing when the Event is triggered.

Loop the Event a few times to test out the flourish. Once you're happy that everything's working, you can change the Probability to lessen how often it plays. Let the Event loop a little while longer, and listen out for repetition. Don't worry if you hear some repeat of combinations – our enjoyment and appreciation of music is based on some degree of repetition and predictability (and the occasional subversion of our expectations) – it only becomes an issue when it starts to stand out and become noticeable.

You can add some more subtle variation by using random volume modulation. Right-click on a volume pot, and choose Add Modulation > Random. You can then control the amount of randomization from the pot that appears on the right-hand side of the deck. Just a few dB on each Instrument can add up to significant changes overall (note: this will not cause constant randomization – the volume is only randomized at the start of each Instrument Trigger Region).

You can use this method on panning, effect parameters and pitch (though changing the pitch of a sample will also affect its tempo).

We could also use parameter control to add variation to the music. We'll come back to this later in the chapter, when we use the time since the Scene was loaded to control the mix levels (and, after a predetermined time, to stop the music).

Nested Probability Loops

Vertical randomization works well, but has some drawbacks – for example, there may be some combinations of Instrument parts that clash when played together.

Another method is to use Horizontal randomization, in the form of nested loops.

Take a look at the Logic Track. You might have spotted that your Tempo Marker is covered by the Loop Region.

- Click on the Loop Region, and drag it upwards, widening the Logic Track.
- Drag GameOverMusicDrumLoop03 from the Audio Bin onto the DrumLoop track, immediately to the right of the existing Multi Instrument.
- Right-click on the new Single Instrument and create a second Loop Region (this will cover GameOverMusicDrumLoop03).
- Right-click in the Logic Track, and create a third Loop Region.
- Drag the third Loop Region ends so that it covers both Instruments in the DrumLoop Audio Track.
- Drag the third Loop Region above both other Loop Regions in the Logic Track.

Play the Event, and see what happens...

Well, that wasn't ideal. The playback cursor is "caught" in the first Loop Region, and never gets to the second one.

To avoid, this, we need to set the first Loop's probability. Select the Loop Region in the Logic Track, and you'll see we can activate the Probability from the Deck. Click on the die button, and try out a Probability of 50%.

This will be slightly better – there's a 50:50 chance of the first Loop working. However, let the Event loop for a while, and you'll see the second Loop never has any effect. Why not?

This is because the vertical order of items in the Logic Track affects how they will behave – the higher they are placed, the higher their priority. As the third Loop Region is at the top, it tells the Timeline Cursor to go to the start (overriding the second Loop Region).

Drag the second Loop Region to the top of the Logic Track, and try again. Now, once you get past the first Loop Region, the Timeline Cursor will remain in the second Loop.

211

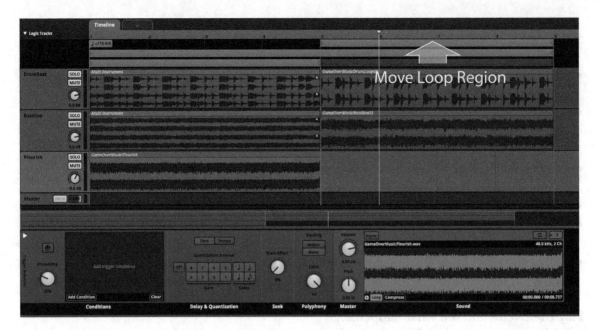

Figure 24.1 Move Loop Region

Finally, select the second Loop Region and change its Probability, so that playback will sometimes go back to the start of the Timeline.

It's worth combining and experimenting with both of these randomization techniques. Together, they will ensure that there is continuous variation to your music.

Parameter Control – Time

Another way to ensure variation to your music is via parameter control – for example, you could use the position of the player in the game to affect the mix. However, as we're in the GameOver menu, this isn't possible. Instead, let's see how we could use the time since the Scene was launched to affect the mix – and after a set time has passed, to stop the music from playing.

Let's start by setting up FMOD. As we're using Loop Regions, we can't use the Timeline – we'll use a parameter value sent from Unity.

- Click on the plus symbol at the top, and select Add Parameter > New Parameter.

- Set it to User Parameter.
- Name it "SceneTime."
- Leave the Minimum value as zero, and set the Maximum to 180.
- Click on OK to close the window (this will take you to the SceneTime parameter).
- Right-click on the DrumLoop volume pot and select "Add Automation."
- Add automation breakpoints to draw in your track automation.
- Repeat the previous two steps for the remaining tracks.

Play the Event, and try changing the SceneTime pot at the top of Event Editor to hear how it will change over time. Don't worry about adding a fade out to the end (we'll deal with this using an AHDSR envelope).

We're done with FMOD for now, so add the GameOverMusic to the GameOverBank, save your project, Build your Banks, and open the GameOver Scene in Unity.

We're going to add the necessary code to the existing GameOver script, so find it in the Console (it should be in the MyScripts folder), and double-click to open it in your IDE.

Grab the code from the GameOverScript2 text file in the book assets, and paste it over so that it looks like this:

213

```
1 using System.Collections;
2 using System.Collections.Generic;
3 using UnityEngine;
4 using UnityEngine.SceneManagement;// allows the script to change the Unity scene
5
6 public class GameOver : MonoBehaviour
7 {
8      private FMOD.Studio.EventInstance gameOverMusic; //GameOver Music FMOD Event
9      private FMOD.Studio.ParameterInstance sceneTime;
10     private float timeCount;
11     public float endGameTime = 180f;
12
13     // Use this for initialization
14     void Start()
15     {
16         FMODUnity.RuntimeManager.PlayOneShot("event:/Player/PlayerImpacts/GameOverImpact"); // play the GameOverImpact Event
17         timeCount = 0;
18         gameOverMusic = FMODUnity.RuntimeManager.CreateInstance("event:/Music/GameOverMusic");
19         gameOverMusic.getParameter("SceneTime", out sceneTime);
20         gameOverMusic.start();
21     }
22
23     // Update is called once per frame
24     void Update()
25     {
26         timeCount += Time.deltaTime;// makes timeCount value increase over time
27         sceneTime.setValue(timeCount);// cast the timeCount value to the FMOD sceneTime parameter value
28         if (timeCount >= endGameTime)
29         {
30             FMOD.Studio.Bus masterBus = FMODUnity.RuntimeManager.GetBus("bus:/");
31             masterBus.stopAllEvents(FMOD.Studio.STOP_MODE.ALLOWFADEOUT);
32         }
33         if (Input.anyKeyDown)
34         {
35             //print("key pressed");
36             if (Input.GetKeyDown(KeyCode.R) || Input.GetButtonDown("Fire1"))
37             {
38                 //print("button pressed");
39                 FMODUnity.RuntimeManager.PlayOneShot("event:/UI_SFX/KeyPressCorrect"); // play the FMOD KeyPressCorrect Event
40                 FMOD.Studio.Bus musicBus = FMODUnity.RuntimeManager.GetBus("bus:/Music");
41                 musicBus.stopAllEvents(FMOD.Studio.STOP_MODE.ALLOWFADEOUT);
42                 SceneManager.LoadScene("Battleground");
43             }
44             else
45             {
46                 //print("wrong key");
47                 FMODUnity.RuntimeManager.PlayOneShot("event:/UI_SFX/KeyPressError"); // play the FMOD KeyPressError Event
48             }
49         }
50     }
51 }
```

Figure 24.2 GameOver script 1

GameOver Script Walkthrough 2

So there's been quite a few changes since we were last here!

We now start by declaring an FMOD Event Instance for the gameOverMusic, an FMOD Parameter instance, and a float, which will be used to hold the time value.

We also declare a public float called "endGameTime," which will be used to set when to stop the music playback.

In the Start Method:

- The timeCount float is set, so that we start counting from zero.
- The gameOverMusic FMOD Event Instance is set so that it triggers the GameOverMusic Event.
- The sceneTime FMOD Parameter Instance is set so that its value is sent to the SceneTime parameter.
- The gameOverMusic FMOD Event is started (which will trigger the GameOverMusic Event).

(Remember that we use lower-case letters at the start of variable names. The variable is named gameOverMusic, and GameOverMusic is the name of the actual FMOD Event. We could have used a totally different name for the variable, but this way you can tell what it's being used for. It may seem confusing at first, but makes sense in practice.)

In the Update Method:

- The timeCount variable is increased by the amount of time taken to complete the last frame – i.e. it acts as a counter.
- The timeCount value is cast to the sceneTime value.
- If the timeCount value is greater than the endGameTime, all FMOD Events are stopped.

Time.deltaTime is a very useful function. Frame rates are not always constant – you must have seen slowdown when a game gets busy (and if not, try adding lots of grass to your Battleground!), so you can't simply add one frame's worth of time to the timeCount value in each Update Method. Time.deltaTime count gives that actual duration that the previous frame took.

+= is a shortcut way of writing "add this to the current value."

Not all operations can be performed on all variable types – for example, we can't use += on an FMOD Parameter Instance. Instead, we perform this on the timeCount float variable, then "cast" the timeCount value to the sceneTime FMOD Parameter Instance.

Stopping all FMOD sounds is performed just like we did in Chapter 20, but this time we use FMODUnity.RuntimeManager. GetBus("bus:/"); to affect all Events going through the Master Bus.

Save your script, and try it out in the game. It should work quite well, but your music is likely to truncate quite abruptly.

- Return to FMOD, and select the GameOverMusic Master Track.
- Right-click on its volume pot, and choose Add Modulation > AHDSR.
- Set the Attack time to zero, and use the Release to set a fade out.
- As ever, save your project and Build your Banks.

If you don't fancy waiting 180 seconds to hear this effect in the game, as the endGameTime variable was set to Public, you can select the Main Camera and change its value from the Unity Inspector. Just remember to change it back once you're finished!

As with randomization, Parameter Automation can be applied to most properties. Why not try adding a low pass filter that gradually removes high frequencies from the mix?

Instruments Inside Parameters

Let's take this opportunity to take another look at using Instruments inside Parameters. Parameters do not have a Timeline – the playback cursor corresponds to the parameter value. Unlike a DAW, the playback cursor does not need to be moving to hear any audio – Instruments are triggered when the cursor crosses their Trigger Region.

- Import the VO_UI_PressRToRestart wav file into your FMOD project, and create a new Audio Track in the GameOverMusic Event.
- Drag the VO_UI_PressRToRestart from the Audio Bin onto the track to create a new Single Instrument (note – this Instrument is to be placed in the SceneTime parameter – not the Timeline!).
- Move this Instrument towards the far right of the track.
- Once again, save your FMOD project, Build your Banks, and play the Scene. You will eventually hear the message "Press R to Restart."

As the FMOD SceneTime parameter was controlled by the time since the Scene was launched, we could have used the Timeline of a separate Event to achieve the same effect. However, this way, hopefully it's now a little clearer how Instruments within parameters behave.

We'll continue to work on music in Chapter 25 – we'll be setting up a score system to control the arrangement.

Game Music Part 2

Parameter Control

Learning Outcomes

By the end of this chapter, you will be able to:

- Use Loop Condition parameter control to arrange music according to the game score.
- Use Transition Markers and parameter control to arrange music according to the game score.
- Use Transition Regions and parameter control to arrange music according to the game score.
- Use Transition Timelines to smoothly change from one music section to another.

In the Chapter 24, we set up music that uses controlled randomization to add variety to prevent noticeable repetition and looping.

But game music can do much more than this. It can be used to inform the player about what's happening – for example, when playing *Zelda: Breath of the Wild*, often the first indication that you've been spotted by an enemy is the change in the music.

In this game, we're going to keep things simple – the score will control the music. There will be several pickups scattered around. Each pickup that is found increases the score. Once the player finds all of these and returns to the Fort, the "YouWin" Scene will load.

Let's start by setting up our music in FMOD. Create a new 2D Event in the Music folder, and label it "BattlegroundMusic." Assign it to a new Bank called "MusicBank," then go to the Mixer window, and assign the Event to the Music Group bus.

Back in the Event Edit window:

- Right-click in the Logic Track to create a Tempo Marker.
- Set the Event tempo to 110bpm.
- Drag the BattlegroundMusicLoop01 file from the Audio Bin onto the Audio Track.
- Position the Single Instrument at the start of the Timeline.
- Right-click on the Single Instrument to set a Loop Region over it.
- Drag the Loop Region above the Tempo Marker (just so you can see clearly what's in the Logic Track).
- Drag over the BattlegroundMusicLoop02 file onto the Audio Track later on the Timeline.
- Position the new Instrument so that it starts on the first bar of beat 3 (immediately after the first loop).
- Right-click on the new Single Instrument (BattlegroundMusic Loop02)and set a Loop Region over it.

As with almost every aspect of game audio, there are a number of approaches we could take here, but all of them will require parameter control.

- Click on the plus symbol and choose Add Parameter > New Parameter.
- Name the parameter "GameScore," and set a range of zero to 6 (this will correspond to the number of pickups in the Scene).
- Click on "OK."

Loop Conditions

Return to the Timeline, and select the first Loop Region. In the Deck you will see there is an option for "Add Condition." Click here and choose Add Parameter Condition > GameScore.

You will now see a green line in the Loop Conditions.

The loop will be active when the GameScore is within the range indicated by the line – at the moment, this is all the time (0 to 6). Change the upper and lower condition limits to 0, and the loop will no longer be active once you score your first point (it can be easier to double-click and type in a value).

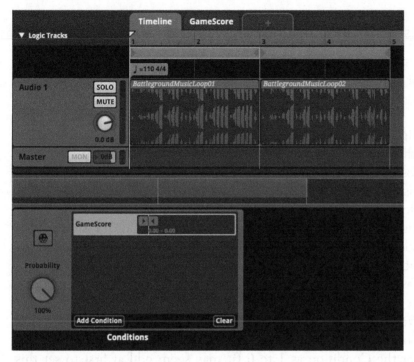

Figure 25.1 Loop Condition

219

Preview this in FMOD. Once the GameScore value is greater than zero, the playback cursor will be released from the first Loop Region (you can double-click on the Parameter value and type it in – FMOD will not respond to the change until you press [Enter] or [Return]).

The two main problems with this technique are:

1) The reaction of the music to the score will be quite slow – if the first loop has just started, you will have to wait quite a while for the change. If you're controlling the arrangement with a different parameter (for example, the distance of the player from the end of the level), then this might work in the game. However, for a scoring system, you probably need a more immediate effect.

2) It is not possible to jump to a totally different part of the Timeline. While this may not be an issue with our scoring system, what if you want to change to a different music section when different conditions are met? (for example, when the player has destroyed all of the enemies, or when they enter a specific part of the level).

Transitions and Destination Markers

A Destination Marker is a form of locate point. When the Playback cursor reaches a Transition, it will cause it to move to a Destination. Let's look at how to combine these with the previous technique.

- Right-click in the Logic Track and select Add Destination Marker (this appears as a gray flag).
- Position the Destination Marker at the start of the second Instrument/Loop Region.
- Double-click on the Destination Marker and name it "Loop2_B1."
- Right-click in the Logic Track and select Add Transition To > Loop2_B1 (this appears as a green flag, pointing towards its Destination).
- Position the Transition halfway through the first Instrument/Loop Region (bar 2, beat 1).
- With the Transition selected, go to the Deck and select Add Condition > Add Parameter Condition > GameScore.
- Set the Condition as 1 to 6 (it may seem odd at first to set this differently to the Loop Region Condition, but remember, we need the Loop Region to be active *before* a point is scored, and the Transition to be active *after* a point is scored).

Preview the Event in FMOD and change the GameScore value before playback reaches the Transition. You will see that when it is active, it will cause the cursor to jump to the start of the second loop

Note – instead of using the Loop Region Condition, you could have used a second Transition set to the same Destination. However, this would need to be positioned at the start of the Timeline, not at the end of the Loop. FMOD currently gives loops priority over Transitions, regardless of the order they are arranged in the Logic Track.

Transition Regions

Let's add a third loop, and use a different method to control playback.

- Drag the BattlegroundMusicLoop03 file from the Audio Bin to the Audio Track.
- Position the Single Instrument at bar 6, beat 1 (it does not have to be adjacent to the second Instrument).
- This is at a different tempo, so add a Tempo marker at the start of the Instrument, and set it to 130bpm.
- Right-click on the Instrument to add a New Loop Region that covers the Instrument in the Timeline.
- Right-click in the Logic Track and select Add Destination Marker.
- Position this at the start of the Loop Region (at the same location as the Tempo marker).
- Name this "Loop3_B1."
- Right-click in the Logic Track over the second Instrument, and choose Add Transition Region To > Loop3_B1. This appears as a green box, with chevrons pointing towards its Destination.
- Position the Transition Region over the second Instrument/Loop Region (BattlegroundMusicLoop02), and drag out the ends to cover it completely.
- With the Loop Region selected, go to the Conditions section in the Deck, and click on Add Conditions > Add Parameter Condition > GameScore.
- Set the Condition range from 2 to 6.

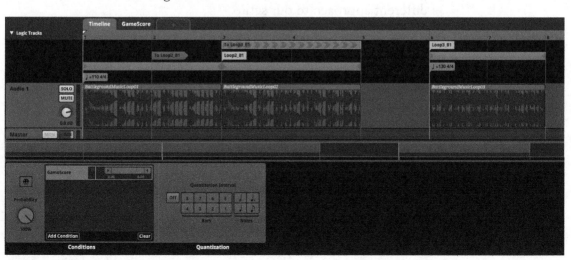

Figure 25.2 Loop Region

Preview the Event in FMOD, and try changing the GameScore value to 2 (or higher) while the Playback cursor is over the Transition Region. You will see that the moment the condition is met, playback jumps straight to the third loop.

This has given us a much more immediate reaction to the parameter change – but unless the player happens to find a pickup exactly on a musical beat, it will result in a sudden rhythmic jump.

This is why Transition Regions offer us a Quantization section. Select the crotchet icon, and you will see green vertical lines representing quarter note intervals. Try this out – you will find that once the condition has been met, the transition will not trigger until the Playback cursor reaches the next green line.

Multiple Transitions and Destination Markers

Transition Regions can be a compromise between reaction time and musicality.

If the music is rhythmic focused, you can get away with abrupt transitions. However, if your music is more melodic, the interruptions in musical phrasing can make the changes quite noticeable.

This is because a Transition Region always jumps to the same destination regardless of the musical beat. A transition from the third beat to the start of a loop might be very obvious to the listener.

Let's go back to the change between the first two loops (BattlegroundMusicLoop01 to BattlegroundMusicLoop02), and take a different approach to our setup.

- Add Destination Markers at beats 2, 3, and 4 of the second loop – (see below).
- Name the Destination Markers accordingly (see below).
- Right-click on the existing Transition (To Loop2_B1) and select "Copy."
- Paste copies of the Transition above the first Instrument at quarter-note intervals (if we copy and paste the Transitions, we don't have to re-configure all of the Conditions – see below).
- Right-click on each of the Transitions and Set Destination To the corresponding beats (see below).

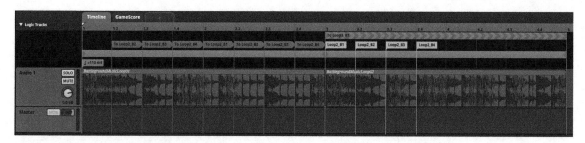

Figure 25.3 Multiple transition markers

This takes longer to set up, but ensures that transitions always go to the corresponding beat of the next loop.

Transition Timelines

Try out the transition to the third loop (BattlegroundMusicLoop02 to BattlegroundMusicLoop03). We've already noted that Transition Regions can cause rhythmic jumps – and this is exacerbated by the change in tempo.

Another way to disguise a change from one musical section to another is to use the Transition Timeline – for example, with a drum fill.

223

- Double-click on the Transition Region, and the Transition Timeline opens up next to it.
- This Timeline will be played before the Destination Marker (these can be added to both Transitions and Transition Regions).
- Drag the BattlegroundMusicFill files into here (creating a Multi Instrument). Now one of these fills will be played before the third loop kicks in.
- Grab and drag the end of the Transition Timeline to set its length – you'll need it to just cover the drum fills.
- Double-click on the Transition Region again to close the Transition Timeline.

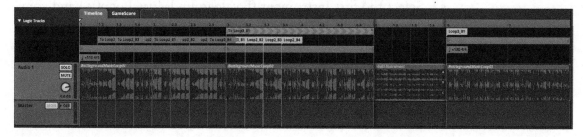

Figure 25.4 Transition Timeline

Be careful! Once you open the Transition Timeline, it will be active – even if you don't put anything in it. This can result in unexpected pauses and silences (especially as you can't see the cursor move while a closed Transition Timeline is playing).

A small circle on the top right indicates if a Transition has a Timeline. Right-clicking on a Transition gives you the option of Remove Transition Timeline, or you can drag the length all the way down to zero.

We didn't actually have to use a Transition Timeline here – we could have positioned the Destination Marker and BattlegroundMusicFill in front of the third loop. However, they can be invaluable when you need to jump into the middle of a loop (for example, they could be added to each Transition over the first loop [bear in mind that you'll have to change the Destinations to maintain the rhythmic flow – e.g. if you have a one-beat Transition Timeline from the second beat, you'll have to set the Destination to beat 3]).

Event Instruments

AHDSR Envelopes are triggered when an Event is started and stopped, not by Instrument Trigger Regions. This means that we can't use them for directly cross-fading Transitions. However, there is a workaround – we can use Events within Events – what FMOD refers to as "nesting."

- Add a new Audio Track in the BattlegroundMusic Event.
- Right-click and select "Add Event Instrument."
- Select the Event Instrument, and double-click on the name at the top right of the Deck to name it "FirstLoop" (you can also find it in the Events Tab, nested under the BattlegroundMusic Event).
- Position the Event Instrument under the first Instrument/Loop Region.
- Adjust its length to match the first Instrument/Loop Region.
- Select the first Instrument (BattlegroundMusicLoop01) and copy it to the clipboard (Edit > Copy).
- Delete the first Instrument (there is no Edit > Cut option in FMOD).
- Double-click on the FirstLoop Event Instrument to open it.
- Paste the first Instrument (BattlegroundMusicLoop01) onto the Event Instrument's Audio Track, and position it at the start of the

Timeline (so don't worry if the pasted Instrument looks shorter than the original – the zoom setting is likely to be set differently in this Event).

- Right-click on the first Instrument to add a covering Loop Region.
- Return to the BattlegroundMusic Event by selecting its name at the top of the Edit window (just under the Transport controls).
- With the FirstLoop Event Instrument selected, right-click on the Volume pot in the Deck and Add Modulation > AHDSR.
- Set the Attack value to zero, and adjust the Release to taste.

Figure 25.5 Nested Instruments

As the first loop is now a nested Event, you will hear it fade out as it transitions to the second Loop Region.

Nested Events are referenced by their "parent" Event, so don't need to be assigned to Banks. As their audio is routed through the "parent" Master Track, you also don't need to worry about Group Bus assignment in the Mixer window.

Unfortunately, envelopes don't work with Transition Regions – these will immediately cut the Event Instruments, regardless of the AHDSR setting. Instead, you will have to pre-render the cross-fades in your DAW as audio files, and place these inside the Transition Region (this also means that you might not be able to use Multi Instruments for your music randomization, as you cannot predict which elements need to be included in the cross-fade).

Occasionally you'll have to work around the limitations of FMOD – but the software is under constant development, and you can expect to see its capabilities expanded and improved with every update.

225

The Design Bible

That's enough to get you started with your music arrangement. Feel free to use the resource loops, but you'll get a lot more from this book if you bring in your own assets.

It's always tempting to throw in "placeholder" music and sound effects (and that's fine when you're first learning how to use software). However, this is often frowned upon, as it's possible for the style and timbre of these to "imprint" on the developers. This leads them to expect a certain character from the audio, so they have preconceptions which affect their perception and judgment when the "real" sounds are added.

Instead, it can be better practice to start off by creating a "design/ style bible." Perhaps the magic spells will all come from nature? Maybe all the enemy weapon sounds should have a common element? What are the melodic themes for each character? What Instruments will be used for the soundtrack? This document will help all of the sound effects and music work together – and it's much easier if these decisions are made early on.

This can be a two-way street – the audio can inspire the design. For example, on Naughty Dog's masterpiece "The Last of Us," because composer Gustavo Santaolalla started work very early in the project, the developers and designers were able to listen to the score while still working on the game. This resulted in a soundtrack and game that match each other perfectly.

Unfortunately, this can't always be the case – budget constraints mean that sound design and music composition often begin quite late into a game's development (though this does have the advantage that you will be working to a much more defined brief).

Take a while to get your music working with the GameScore parameter, then save your FMOD project and Build your Banks. I've kept things quite simple, but feel free to see how much further you can take your own game.

In Chapter 26, we'll be setting up the code and Unity side of our music system.

226

Setting Up Unity for Music

Learning Outcomes

By the end of this chapter, you will be able to:

- Configure Unity with the Music System script.
- Modify the Reverb zone to prevent reverb still being applied to the mix when the game restarts.

Pickups

Let's start off with our pickups. You could use a 3D model for these (follow the same procedure we used for importing the rocket launcher, Sentry Drones and health pickups), but to keep things simple, I'm just going to use Unity's sphere primitive.

- Go to GameObject > 3D Object > Sphere.
- Name this "pickup."
- Scale and position the pickup so that the player will be able to reach it in the Scene.
- Go to Edit > Project Settings > Tags and Layers, and create a new Tag called "Pickup" (you can also get to this from the Tag menu at the top of the Inspector).

- Create another Tag called "WinTrigger" (we'll need this later on).
- Tag the pickup as "Pickup" (it would be possible to use the name of a GameObject in our code instead of Tags. However, Tags are more efficient, so it's better to get into the habit of using them where possible).
- Set the Sphere Collider component to "Is Trigger."

To make the pickup more interesting looking, go to the MyMaterials folder, right-click in the gray-space of the Project Tab and create a new Material. Name this "PickupMaterial," adjust the component values in the Inspector to taste, before dragging the Material onto the pickup GameObject.

The player wins the game once they have found all the pickups in the Scene and returned to the Fort. This means that we'll need a collider in the Fort to trigger the Win state. Let's use the existing collider for the Reverb Zone, so select this in the Hierarchy (FortReverbZone – it should be a child of the Fort), and assign it the "WinTrigger" Tag (this also has the advantage that it is already in the NotRocketLayer layer, so it will not interact with the Rockets).

Now navigate to the MyScripts folder, right-click in the Project Tab gray-space, and create a new C# script called "MusicSystem." Drag this onto the FPSController, then double-click to open it in your IDE.

As ever, you'll find the code for the script with the book resources (MusicSystem1.txt).

```
1 using System.Collections;
2 using System.Collections.Generic;
3 using UnityEngine;
4 using UnityEngine.SceneManagement;
5
6 public class MusicSystem : MonoBehaviour
7 {
8     public FMOD.Studio.EventInstance battlegroundMusic; //Music FMOD
9     private FMOD.Studio.ParameterInstance gameScore; //PointSystem for FMOD Parameter
10    private int countScore; //Score
11    private int maxScoreCount;
12    public float winTriggerPause = 2f;
13
14    // Use this for initialization
15    void Start()
16    {
17        countScore = 0;  // Starting score
18        maxScoreCount = GameObject.FindGameObjectsWithTag("Pickup").Length;// counts the number of pickups in the scene
19        battlegroundMusic = FMODUnity.RuntimeManager.CreateInstance("event:/Music/BattlegroundMusic"); // Makes the Music event trigger the FMOD Event
20        battlegroundMusic.start(); //Play music on game start
21        battlegroundMusic.getParameter("GameScore", out gameScore);
22        //SayMyName("Ciaran");
23    }
24
25    void OnTriggerEnter(Collider other)
26    {
27        if (other.gameObject.tag == "Pickup")//Determines tag of objects that will cause score to increase
28        {
29            countScore = countScore + 1; //Increase countScore by 1
30            gameScore.setValue(countScore);
31            //print ("Score = " + countScore);
32            Destroy(other.gameObject); //Destroy the collided with GameObject
33        }
34        if ((other.gameObject.tag == "WinTrigger") && (maxScoreCount <= countScore))
35        {
36            Invoke("TriggerYouWin", winTriggerPause);
37            //print ("Call YouWin");
38        }
39    }
40    void TriggerYouWin()
41    {
42        //print ("Yay!");
43        SceneManager.LoadScene("YouWin");
44    }
45
46    // void SayMyName (string myNameHere)
47    //{
48    //      print ("my name is " + myNameHere);
49    //}
50 }
```

Figure 26.1 MusicSystem script 1

Code Walkthrough: Music System

This code is again using the SceneManagement class, as it needs to be able to change to the "YouWin" Scene.

We then declare a number of variables:

- battlegroundMusic: an FMOD Event Instance which will trigger the music to play.
- gameScore: an FMOD parameter which will control the music arrangement.
- countScore: this integer will keep count of the score value, which will then be passed to the gameScore FMOD parameter.
- maxScoreCount: will count how many Pickups are in the Scene when it starts.

229

- winTriggerPause: sets a delay between winning the game and loading the YouWin Scene.

Start Method

When the Scene launches:

- The countScore value is reset to zero.
- The number of GameObjects with the tag "Pickup" are counted, and this value is assigned to the maxScoreCount integer.
- The battlegroundMusic Event Instance is set to the FMOD BattlegroundMusic Event.
- The battlegroundMusic Event Instance is triggered.
- The gameScore Parameter Instance is assigned to the FMOD BattlegroundMusic Event GameScore parameter.
- (Ignore the commented-out "SayMyName" function call for now – we'll look at this in a moment.)

OnTriggerEnter Method

This has two sets of conditions.

If the player collides with a GameObject tagged as "Pickup":

- The countScore value is increased by one.
- The countScore value is passed to the gameScore.
- The pickup is destroyed.

If the player collides with a GameObject tagged as "WinTrigger" and the countScore value is equal to the maxScoreCount (the number of Pickups initially in the Scene):

- The TriggerYouWin function is invoked.

When a function is invoked, you have the ability to set a delay before it is actioned. This helps in the game, as otherwise the YouWin Scene is launched the moment the player enters the Fort (which seemed too abrupt when I play-tested the level).

Brackets after a function name have a purpose – they allow you to set values within the function. To demonstrate this, I've put a function called SayMyName on lines 46 to 49, and called it inside the Start Method (on line 22).

Line 46 declares the function, and also says that it has a string (text value) inside called myNameHere.

When I call the function, if I put a text string inside the brackets, this will be passed inside the function, into the place of myNameHere.

So, if you un-comment lines 46 to 49 and line 22, when the game is played, "my name is Ciaran" is printed to the Console.

Well, this isn't particularly useful, so you might as well comment it out again. However, you should now be able to make more sense of what happened at line 31.

Go to Help > Scripting Reference, and search for "Invoke." This will give you quite a few matches, but we're looking for MonoBehaviour. Invoke. Select this, and you can now see the entry for the Invoke function.

- It tells you that it's public – well, we knew that – otherwise we wouldn't be able to access it from our script.
- Its type is void – i.e. it does not give an output value.
- Two values can go inside the brackets (these will be separated by a comma):
  - The name of the method/function that is to be invoked (the name value is entered as text, i.e. as a string).
  - The delay time before the method/function is invoked (as a float value).

Line 31 therefore means that the TriggerYouWin function is invoked after a delay. The delay value is set from the winTriggerPause float (which is a public variable, and therefore can be accessed from the Unity Inspector).

The TriggerYouWin function simply loads the YouWin Scene.

Problems with the Reverb Zone

Play the game, find the pickup and return to the Fort. Press [R] to restart, and you might notice a bug. If your player starts off outside the Fort, you'll hear reverb. This is because you triggered the Reverb Mix Snapshot when you entered the Fort. As re-starting the game immediately moves you outside the Fort, Unity never detected you exit

231

the collider to stop the Event. We're going to have to take a different approach to triggering the reverb, and we'll take advantage of the way that FMOD can include Snapshots inside Events.

We'll start by setting up FMOD.

- Create a new 2D Event in FMOD and name it "FortReverbZoneEvent."
- Assign it to the BattlegroundBank.
- Right-click in the Audio Track and choose Add Snapshot Instrument > InFort.
- Position the Snapshot Instrument at the start of the Timeline.
- Right-click on the Snapshot Instrument and add a Loop Region that covers the Instrument Trigger Region.
- In the Mixer window, assign the FortReverbZoneEvent to a new Group called "ReverbControl."
- Save the FMOD project and Build the Banks.

Now we'll change the way that Unity triggers the Reverb Zone. Select the FortReverbZone GameObject, and remove the FMOD Studio Emitter (use the cog icon at the top right of the component).

Next, navigate to the MyScripts folder, and create a new C# script called "ReverbZone." Attach this to the FortReverbZone, then open it in your IDE, paste in the code from the ReverbZoneScript.txt resource, and save the script.

```
1 using System.Collections;
2 using System.Collections.Generic;
3 using UnityEngine;
4
5 public class ReverbZone : MonoBehaviour
6 {
7     public FMOD.Studio.EventInstance fortReverbZoneEvent;
8     //The FMOD Snapshot (InFort) is contained within a looped Snapshot Instrument on the FortReverbZoneEvent.
9     //Route the FortReverbZoneEvent in the FMOD Mixer to a bus called "ReverbControl".
10    //  This allows you to stop all instances of the Event with one statement.
11
12    void stopReverb()
13    {
14        FMOD.Studio.Bus reverbControl = FMODUnity.RuntimeManager.GetBus("bus:/ReverbControl");
15        reverbControl.stopAllEvents(FMOD.Studio.STOP_MODE.ALLOWFADEOUT);
16    }
17
18    void Start()
19    {
20        fortReverbZoneEvent = FMODUnity.RuntimeManager.CreateInstance("Event:/FortReverbZoneEvent");
21        stopReverb();
22    }
23
24    void OnTriggerEnter(Collider other)
25    {
26        if (other.gameObject.tag == "Player")
27        {
28            fortReverbZoneEvent.start();  //Trigger the FortReverbZoneEvent when the Player enters the collider.
29        }
30    }
31    void OnTriggerExit(Collider other)
32    {
33        if (other.gameObject.tag == "Player")
34        {
35            stopReverb();  //stop the FortReverbZoneEvent when the Player exits the collider.
36        }
37    }
38 }
```

Figure 26.2 ReverbZone script

ReverbZone Script Walkthrough

Because of the way that FMOD deals with snapshots, this is actually quite a difficult problem to solve.

The obvious method to use is called OnTriggerStay, which checks to see if a GameObject remains inside the collider. Unfortunately, this can be a bit temperamental, and testing found that it only triggers 50% of the time (this is because its frequency is controlled by the physics timer, not the frame rate).

We could try to fix this by tweaking the time settings and playing with the way that the method is called, but as OnTriggerStay is continuously checking for collisions, it's always going to be eating up some of your processing power.

OnTriggerEnter and OnTriggerExit are more reliable and efficient, but we still have the same problem as before (when the game restarts, you'll still hear the reverb).

We can't simply use a stop or release statement (such as reverbZoneEvent.stop(FMOD.Studio.STOP_MODE. ALLOWFADEOUT);), as we need to stop the specific instance of the FMOD Event. However, we have routed the Event though a bus, and (as we've seen before) we have the ability to stop all Events going through a single bus.

We only have one variable to declare: the fortReverbZoneEvent FMOD Event Instance.

Next, the stopReverb function is declared (we could have placed this anywhere in the code – the only reason I have moved this above the Start Method is to help you to follow the code explanation).

When this is called:

- We declare a bus called reverbControl, and set this to affect the FMOD bus called ReverbControl.
- We stop all Events passing through this bus.

In the Start Method, we set the FMOD Event Instance to trigger the FortReverbZoneEvent FMOD Event, and trigger the stopReverb function.

We then set up the OnTriggerEnter Method so that it will trigger the fortReverbZoneEvent FMOD Event Instance if a GameObject tagged as "Player" enters the collider of the GameObject that the script is attached to.

Finally, we set up the OnTriggerExit Method so that it will trigger the stopReverb function if a GameObject tagged as "Player" leaves the collider of the GameObject that the script is attached to.

This took quite a while to get working! Perhaps the solution isn't particularly elegant, but sometimes you have to think outside the box to get things to behave the way that you need.

When you take an unconventional approach to coding, it's really important that you use comments to document what (and why) you did – otherwise it will be next to impossible to come back at a later date and make sense of everything.

Playtesting

As jobs in game development go, playtesting isn't the most glamorous (or well paid), but it's extremely important. Somewhat ironically, this isn't all fun and games. It involves methodically testing every combination and sequence of possibilities, trying to break it. Once any issues are discovered (and documented), playtesters then have to try and reproduce the problems, and narrow down exactly under what conditions they occur, so that the developers can figure out an approach to fix them.

Once bugs have been found, they're triaged according to severity and importance to fix. Game-breaking problems will have to be fixed, but minor and rarely occurring bugs may be left until post-release, or never fixed at all (as anyone who's played *Skyrim* can testify…).

As most games consoles and PCs are connected to the internet, we've seen the rise of the "day zero patch" – software updates and bug fixes, released on the day of the game's release. This allows developers to continue working on projects even after the game has gone to "gold master" (the version of the game sent for production and release).

Every year we hear about buggy games being released, some of which are reported as almost unplayable (*Batman: Arkham Knight* and *Mass Effect: Andromeda* being a couple of notable examples). While this is partly due to commercial pressure to release games on schedule, you should remember that as games get more and more complex, the potential for bugs also increases.

Hopefully your game will be free from bugs, but your levels will still need playtesting. Can the player fall off the game world? Does the environment lead the player though the game, or will they wander for ages trying to figure out the objectives? Think about the purpose of the game. If it is intended as a portfolio of your work, playtesting can ensure that it's not only fun, but that people can easily see everything you're capable of creating.

Navigate to the MyPrefabs folder, and drag over the pickup from the Hierarchy. This gives you a pickup prefab, which you can drag as many times as you need into the Scene.

Remember that we set the GameScore Parameter range from 0 to 6, so if you wish to add more pickups than this, you may also want to add more interactive music.

The game is almost done! In Chapter 27, we'll be putting the finishing touches by adding a score display, sound effects for when pickups are found, and looking at Sidechaining techniques.

235

Mixing the Game

Sidechains

236

Learning Outcomes

By the end of this chapter, you will be able to:

- Modify Unity and the MusicSystem script to display the game score.
- Modify FMOD and the MusicSystem script to add a sequential audio score count.

At the moment, aside from the music changes, there is no indication that you have scored a point. Let's fix that by adding a display to show the score, as well as adding a sound effect.

The score is based on the Pickup count, so rather than start from scratch, we'll modify the MusicSystem script, so open this up in your IDE. You can paste over the code from the MusicSystemScript2. txt file:

```
 1 using System.Collections;
 2 using System.Collections.Generic;
 3 using UnityEngine;
 4 using UnityEngine.SceneManagement;
 5 using UnityEngine.UI;
 6
 7 public class MusicSystem : MonoBehaviour
 8 {
 9     public FMOD.Studio.EventInstance battlegroundMusic; //Music FMOD
10     private FMOD.Studio.ParameterInstance gameScore; //PointSystem for FMOD Parameter
11     //public FMOD.Studio.EventInstance speakCount; //FMOD Event to announce the score
12     private int countScore; //Score
13     private int maxScoreCount;
14     public float winTriggerPause = 2f;
15     public Text showScore;
16
17     // Use this for initialization
18     void Start()
19     {
20         countScore = 0;  // Starting score
21         gameScore.setValue(countScore);
22         maxScoreCount = GameObject.FindGameObjectsWithTag("Pickup").Length;// counts the number of pickups in the scene
23         showScore.text = countScore.ToString() + "/" + maxScoreCount.ToString();
24         battlegroundMusic = FMODUnity.RuntimeManager.CreateInstance("event:/Music/BattlegroundMusic"); // Makes the Music event trigger the FMOD Event
25         battlegroundMusic.start(); //Play music on game start
26         //speakCount = FMODUnity.RuntimeManager.CreateInstance("event:/UI_SFX/SpeakCount");
27         battlegroundMusic.getParameter("GameScore", out gameScore);
28         //SayMyName("Ciaran");
29     }
30
31     void OnTriggerEnter(Collider other)
32     {
33         if (other.gameObject.tag == "Pickup")
34         {
35             countScore = countScore + 1; //Increase countScore by 1
36             gameScore.setValue(countScore);
37             showScore.text = countScore.ToString() + "/" + maxScoreCount.ToString();
38             //if (countScore == 1)
39             //{
40             //    speakCount.start();
41             //}
42             //else
43             //{
44             //    speakCount.triggerCue();
45             //}
46             print("Score = " + countScore);
47             Destroy(other.gameObject); //Destroy the collided with GameObject
48         }
49         if ((other.gameObject.tag == "WinTrigger") && (maxScoreCount <= countScore))
50         {
51             Invoke("TriggerYouWin", winTriggerPause);
52             //print ("Call YouWin");
53         }
54     }
55     void TriggerYouWin()
56     {
57         //print ("Yay!");
58         SceneManager.LoadScene("YouWin");
59     }
60
61     //void SayMyName (string myNameHere)
62     //{
63     //    print ("my name is " + myNameHere);
64     //}
65 }
```

Figure 27.1 MusicSystem script 2

Modifying the MusicSystem to Display the Score

We need the UI Class to be able to display the score, so we've added "using UnityEngine.UI;" at line 5.

We then declare a public Text variable (showScore), which will allow us to display the score as a text string.

In the Start Method, we insert the line of code:

```
showScore.text = countScore.ToString() + "/" + maxScoreCount.
ToString();
```

This sets the showScore variable to display the current score, followed by a forward-slash, followed by the maximum score value (the ToString command converts the countScore and maxScoreCount integer values to strings, so that they can be placed in the showScore text variable).

The same line of code is also placed in the OnTriggerEnter Method, so that the score display will update whenever a point is scored.

Save the script, return to Unity and go to GameObject > UI > Text to add a Canvas and a Text GameObject to the Scene. Name the Text "ScoreText."

So that you can see how the Text will be displayed in the game, change from the Scene Tab to the Game Tab.

(You should be able to see the Text, but depending upon where it was placed, it might be hidden. Don't worry – we're about to lock the text to the top left of the screen, which will also bring it into view.)

Select the FPSController, and look for the MusicSystem component in the Inspector. You'll see a box labeled "Show Score." Drag the ScoreText from the Hierarchy into this box (this sets which Text GameObject the script Text variable affects).

At the moment, the ScoreText position is set relative to the center of the screen. At times this can be useful, but it can look strange when the screen resolution is changed (or if the window is resized). Instead, we're going to lock the text to the top left of the game screen, so:

• Select the ScoreText.
• Option/alt-click on the RectTransform box in the Inspector, and select the left-top box.

This will place the text right in the corner, so adjust this to taste using the X and Y position Transform values (the easiest way to do this is by clicking and dragging on where it says "Pos X" and "Pos Y" in the Inspector component).

By default, the only available font is Arial. To add more options, import a.ttf (TrueType) or.otf (OpenType) font file into your Unity project assets folder. You probably have several of these installed on your computer, and you can find hundreds more online.

Once you've imported a font file:

- Select the ScoreText GameObject.
- Click on the icon of a dot inside a circle just to the right of the Font name in the Inspector to open the Select Font menu.
- Double-click to choose your font.

You can then set the font size, alignment, and color in the Inspector. You might find that the text seems to disappear if you make the font too large – this is because it's too big to fit in the text box. You can fix this by altering the Width and Height settings (in the Rect Transform component), or by changing the Horizontal and Vertical Overflow from Wrap to Overflow (in the Text component).

Now let's add a sound effect to play whenever a point is scored. There are several ways that we could do this:

FMOD Event Emitter

Attach an FMOD Event Emitter to the pickups to trigger OnCollision. This will be fine if we simply want to play the same sound effect every time.

Multiple FMOD Events

Our score point effect is going to be a voice-over that states the number of points scored. We could write a script that counts the number of points scored, and trigger a different Event for every point value. This will be effective, but using multiple Events might make working in FMOD quite awkward.

Sustain Points

These are "markers" in the Logic Track that pause the Timeline. Playback will continue once a "Cue" is triggered on the Event. These can be extremely useful in FMOD, so let's see how to use these for our score sound.

We'll start by setting up our FMOD Event.

- Create a new 2D FMOD Event called "SpeakCount" in the UI_ SFX folder.
- Assign this to the PlayerBank.
- Import the VO_UI_SpeakCount wav files to your FMOD project.
- Drag these from the Audio Bin to create Single Instruments on the SpeakCount Audio Track.
- Arrange these sequentially, so that when you play the Event, you hear the numbers count from one to six (see below).
- Right-click in the Logic Track, and choose "Add Sustain Point."
- Position the Sustain Point at the start of the VO_UI_SpeakCount_ 02 Instrument.
- Repeat the last two steps for the remaining Instruments.
- In the FMOD Mixer window, route the SpeakCount Event to the UI Group (either right-click and choose Reroute Into > SFX > SFX_NoHitFilter > UI, or click and drag it to the Group).
- Save the FMOD project and Build the Banks.

Figure 27.2 Sustain markers

If you preview the Event in FMOD, you'll see that the Timeline cursor stops at the Sustain point. Pressing the "Cue A" button at the top will resume playback (there are no Cues B to Z in FMOD – only Cue A. It's likely that more will be added at a later date).

That's FMOD all good to go – now we just need to modify our MusicSystem script again. This necessary code was already included earlier in this chapter – all you have to do is to un-comment the appropriate lines.

- Remove the forward-slashes at the start of line 11 (public FMOD. Studio.EventInstance speakCount;).
- Remove the forward-slashes at the start of line 26 (speakCount = FMODUnity.RuntimeManager.CreateInstance("event:/UI_SFX/ SpeakCount");).
- Remove the forward-slashes at the start of lines 38 to 43 (don't forget that you can do this in one go by making a selection across all the lines, and pressing [cmd]+[/] (Mac) or [ctrl]+[/] (Windows)).

```
1 using System.Collections;
2 using System.Collections.Generic;
3 using UnityEngine;
4 using UnityEngine.SceneManagement;
5 using UnityEngine.UI;
6
7 public class MusicSystem : MonoBehaviour
8 {
9     public FMOD.Studio.EventInstance battlegroundMusic; //Music FMOD
10     private FMOD.Studio.ParameterInstance gameScore; //PointSystem for FMOD Parameter
11     public FMOD.Studio.EventInstance speakCount; //FMOD Event to announce the score
12     private int countScore; //Score
13     private int maxScoreCount;
14     public float winTriggerPause = 2f;
15     public Text showScore;
16
17     // Use this for initialization
18     void Start()
19     {
20         countScore = 0;  // Starting score
21         gameScore.setValue(countScore);
22         maxScoreCount = GameObject.FindGameObjectsWithTag("Pickup").Length;// counts the number of pickups in the scene
23         showScore.text = countScore.ToString() + "/" + maxScoreCount.ToString();
24         battlegroundMusic = FMODUnity.RuntimeManager.CreateInstance("event:/Music/BattlegroundMusic"); // Makes the Music event trigger the FMOD Event
25         battlegroundMusic.start(); //Play music on game start
26         speakCount = FMODUnity.RuntimeManager.CreateInstance("event:/UI_SFX/SpeakCount");
27         battlegroundMusic.getParameter("GameScore", out gameScore);
28         //SayMyName("Ciaran");
29     }
30
31     void OnTriggerEnter(Collider other)
32     {
33         if (other.gameObject.tag == "Pickup")
34         {
35             countScore = countScore + 1; //Increase countScore by 1
36             gameScore.setValue(countScore);
37             showScore.text = countScore.ToString() + "/" + maxScoreCount.ToString();
38             if (countScore == 1)
39             {
40                 speakCount.start();
41             }
42             else
43             {
44                 speakCount.triggerCue();
45             }
46             //print ("Score = " + countScore);
47             Destroy(other.gameObject); //Destroy the collided with GameObject
48         }
49         if ((other.gameObject.tag == "WinTrigger") && (maxScoreCount <= countScore))
50         {
51             Invoke("TriggerYouWin", winTriggerPause);
52             //print ("Call YouWin");
53         }
54     }
55     void TriggerYouWin()
56     {
57         //print ("Yay!");
58         SceneManager.LoadScene("YouWin");
59     }
60
61     //void SayMyName (string myNameHere)
62     //{
63     //    print ("my name is " + myNameHere);
64     //}
65 }
```

Figure 27.3 MusicSystem script 3

> ## Modifying the MusicSystem Walkthrough 2
>
> The speakCount FMOD Event Instance is treated exactly the same way as we've seen before. First it's declared (line 11), and then set to trigger the appropriate FMOD Event (line 25).
>
> In the OnTriggerEnter Method, we add to the existing IF condition (), so that:
>
> - If the Collided with object is tagged as "Pickup," if the countScore value is equal to 1, then the speakCount Event Instance is triggered.
> - If the Collided with object is tagged as "Pickup," if the countScore value is not equal to 1 (else), then the Cue on the speakCount Event Instance is triggered.

Sidechaining

Sidechaining is where the level of an audio signal is used to control another parameter. The most obvious example of this is when one signal lowers the level (or "ducks") another.

This will help our mix – at the moment, when the ScoreCount Event plays, it can be difficult to hear this behind the music.

In the FMOD Mixer window, select the Music Group, and add a Compressor effect. Turn on the Sidechain option (so that we can definitely hear the ducking happen), set the ratio to around 10:1 (this sets the amount of the effect), and lower the Threshold to -35dB (this sets the minimum signal level that will cause gain reduction – see Chapter 18 for a more detailed look at compressor parameters.

Next, select the UI Group, and click on the plus symbol in the Deck (a pre-fader sidechain feed will be easier to configure in the mix, so use the plus symbol to the left of the Volume pot). Choose Add Sidechain > Music:Compressor, and (once you Save and Build), you're ready to try it out and experiment with the Compressor parameters. It is possible to open multiple FMOD Event windows to play the Music and the ScoreCount Events at the same time, but you might find that this is an ideal opportunity to use Live Update (see Chapter 20).

If you wish to only have the ScoreCount Event duck the music (and not the other UI sounds), then this is possible – remove the existing

Sidechain, and select the UI_SFX_SpeakCount in the Routing Tab to view its settings in the Deck. Click on a plus symbol to Add Sidechain. Note that you cannot directly set the Sidechain destination from here, but if you right-click on the Sidechain again, you now have the option of Connect To Effect > Music:Compressor.

Ducking is a common use of sidechains, but FMOD can do much more. Almost any parameter can be controlled via a sidechain.

Let's try out something strange: open the SentryDroneNoises Event. This consists of a looping humm, and occasional spark sound effects. Let's make the buzzes change the pitch of the humm.

- Select the Humm Single Instrument to view its properties in the Deck.
- Right-click on the Pitch pot and select Add Modulation > Sidechain.
- Select the Buzz Audio Track to view its properties in the Deck.
- Click on a plus symbol and choose Add Sidechain > New Sidechain.
- Right-click on the Sidechain, and choose Connect To Modulator > Humm:SFX_DroneHumm_Loop01:Pitch.
- Re-select the Humm Single Instrument.

243

You now have a section of the Deck labeled Sidechain:Pitch. From here, you can set exactly how it behaves. Play around with this until you're happy with the way that it sounds. If you don't like the effect, get rid of it! This was really just an opportunity to see what else can be achieved using FMOD Sidechains.

Our game is now complete! Well – that's all we're going to be covering in this book; obviously you can take yours much further. For now though, it's time for you to work on your level design, and to get everything playable. If you're planning to use this as a portfolio piece, then here are a few tips:

- Keep the level compact – you need the player to quickly be able to see all of your work.
- Don't make it too difficult! – though there should be some element of a challenge, you don't want the player to give up before they make it to the end of the level.
- Signpost the direction the player is to go. The route through the level should be obvious, and the player shouldn't have to

back-track to find anything they missed. This can be done through level design (for example, by setting out paths in the Terrain) and audio cues.

- Maintain a consistency to your sound design – all elements should come from the same sonic palette to ensure they work together.
- Playtest your level! Then playtest it again! A broken game won't impress anyone – and mistakes such as spelling errors show that you don't pay enough attention to detail.
- Aim for the finished product to run on a medium-spec computer – you never know what will be used to play your game. We'll look at audio optimization in Chapter 28, but avoid using over-complex meshes, and remember that adding grass to your Terrain can quickly slow down the frame rate.
- Keep it simple! You're not looking to show off your game design skills (at least, that's not the aim of this book). It should be immediately obvious to the player what they need to do – perhaps you could add a splash-screen that shows the game objectives and controls? (you could use a modified version of the YouWin Scene).

244

Work on this for a while, and once you're ready, move on to Chapter 28, where we'll be looking at optimizing the performance of your audio.

Optimizing the Game

Learning Outcomes

By the end of this chapter, you will be able to:

- Use the FMOD Profiler to meter the performance of the Audio Middleware.
- Use voice stealing to limit and control the number of Events in the game.
- Load and unload FMOD Banks as and when they are needed by the game.
- Set and configure the data compression type for individual audio resources.

Audio processing is rarely the greatest drain on a game's resources – this is far more likely to be due to the lighting, graphics, shaders, and physics calculations. However, if we can optimize the audio perform-ance, this frees up processing power for everything else.

To help with this, FMOD includes the Profiler. You'll find this under Window > Profiler.

This allows us to see and record exactly what the audio is doing as the game is played.

Make sure that Live Update is configured (see Chapter 20), and start your game playing in Unity. While it's playing, change over to FMOD and hit the Record button in the Transport. Play the game for a while. When you stop the game, the Profiler should also stop recording.

FMOD will have created a new Profiler Session (called "New Session"), which consists of multiple tracks – one for each Event triggered in the game, and the Master Bus.

If there's too much here at one time to be able to tell what's going on, right-clicking on a track header gives you the option of Remove Track (and Remove All Tracks).

You can add Buses to the display by choosing Add Tracks > Recorded Buses. There's also an option to show tracks for Events and Groups included in a custom mixer view (e.g. Add Tracks > From Mixer View > Player).

The Master Bus has a disclosure arrow that allows you to see even more information about the performance.

Press Play, and you should hear your recording (if you can't hear anything, try changing the Playback with the API Capture button [labeled API, next to the Transport position display] setting).

Pay attention to the Overview on the right, and you'll be able to see the 3D position of Events relative to the listener, as well as seeing when Snapshots are triggered.

API

API stands for Application Programming Interface. In this case, it lists the functions that are called from FMOD.

Near the Transport, there is a button that changes the Profiler view from Tracks view to API Capture. Change to the API view, and you'll see line upon line of text. Take a closer look at this, and you might be able to make sense of some of it – the messages are similar to the code we've used.

API view can be useful for debugging (figuring out exactly when and what causes issues with the sound), but it's beyond the scope of what's covered in this book.

At the top of the Profiler window are a number of radio buttons (only one can be selected at a time). These allow you to switch between different performance displays:

- CPU: shows the percentage of CPU power used by an Event. Useful for seeing which Events are using up your processing power. You're likely to see this spike when effects are added.
- Memory (alternative to File I/O): only available on the Master Bus. Shows the RAM used, and is displayed in blue.
- File I/O (alternative to Memory): only available on the Master Bus. Shows the amount of data loading to RAM (shown in green) and streaming from disk (shown in purple).
- Levels: shows the signal levels of Events and buses.
- Voices: shows the number of voices used on a track/bus. "Self" will show the voice count of an Event, and "Total" also includes any nested or referenced Events.
- Lifespans: shows a line that displays the duration of each Event instance.
- Instances: shows the number of instances of an Event. Also can change between "Self" and "Total" (see Voices).

To see the exact value of any of these at a point in time, moving your mouse over the graph will show the position and value at the cursor point.

So now that you can see how the game is performing, what can we do to make this more efficient?

Voice Count

The fewer voices, the more efficient the game audio. Let's take a look at the SentryDroneNoises Event, and see if we could lower its voice count.

Remember that this used three tracks – Humm, Buzz, and a Scan sound effect. If you recorded a Profiler Session that includes a SentryDrone Event, you'd see the voice count change whenever a spark is played.

Select the Buzz Scatterer Instrument, and take a look at its settings in the Deck. We could lower the Polyphony to 2, which will ensure that no more than two sparks will play at a time. (Don't get Polyphony mixed up with the Total Sounds control – this counts the number of sounds triggered, and will stop once that count is reached. For example, if I set this to 5, the Scatterer Instrument stops making sparks after it's played five of them.)

Select the Master Track, and take a look at the far right-hand side for the parameter labeled "Max Instances." This determines how many Instances of an Event can be played at the same time. Change this to 2, and only two of your SentryDrones will now make a sound.

If you try to play any more instances of the Event, then one of them will stop (i.e. its voice is "stolen"). You can determine which Events have their voices stolen by changing the Stealing parameter:

- Oldest: the oldest Event instance will stop.
- Furthest: the furthest Event instance will stop.
- Quietest: the quietest Event instance will stop.
- Virtualize: the quietest Event is Virtualized. This means that it is silenced until it would no longer be the quietest instance.
- None: no new Event Instances can be played until an existing Instance is stopped.

Virtualize is the only option here that is dynamic – the other settings are only relevant when we try to play a new Event instance. Our SentryDroneNoises Events are all triggered when the game starts, so if (for example) we set this to Furthest, then only the two Drones closest to the player will make any sound. Shooting these Drones or moving away from them won't cause the other Drones to now emit sound, as nothing is configured to try to trigger the Event instances again.

But Virtualize works differently, and steals and returns voices according to the levels of the Event Instances (which, as they are 3D Events, will be affected by their distance from the player) – exactly what we need.

Figure 28.1 Virtualize setting

249

You'll have to experiment with the Max Instances value – it will depend on your game level, and how far apart you've spaced the SentryDrones – you might even be able to get away with setting this to 1.

Bank Loading

By default, FMOD will load all of your Banks along with the Scene. That's useful when you're putting your game together, but there's no advantage to loading all the sounds for all your levels all the time.

In Unity, go to FMOD > Edit Settings, and uncheck the Loading: Load All Event Data at Initialization (this label might be truncated – try changing the size of the Inspector to see the full text). If you now press play, you'll see quite a lot of error messages as the game tries to play unloaded Events.

Add a new Empty GameObject to the Scene, name this "BanksLoader," and select it in the Hierarchy.

Which Banks do we need to load? For the Battleground Scene we'll definitely need:

- Master Bank: we've avoided using the Master Bank for our Events. This is because it's used for the mixer settings, and therefore is needed in all of our Scenes.
- BattlegroundBank: obviously! This includes the CrateImpact Events, the explosion and weapon Events, as well as the Ambience and fortReverbZoneEvents.
- PlayerBank: contains all the player foley and SpeakCount VO Events.
- MusicBank: contains the BattlegroundMusic Event (not the GameOverMusic Event).
- SentryBank: contains the Events for the SentryDrone.

We'll want all of these to load with the Scene. However, we can unload the PlayerBank, the SentryBank and BattlegroundBank Banks when the level is finished.

We'll keep the MainBank and MusicBank loaded, as we will need elements of these in our other Scenes. Therefore we'll need two different FMOD Bank loaders.

- Click on Add Component, and add FMOD Studio > FMOD Studio Bank Loader.
- Click on the Add Bank button.
- Choose the Master Bank from the pop-up menu.
- Repeat the process to add the MusicBank.
- Set the Load menu to "Object Start."
- Leave Unload set to "None."

Now we'll need a second FMOD Bank Loader. Set this up in exactly the same way (but add the PlayerBank, SentryBank and BattlegroundBank Banks), and set the Unload menu to "Object Destroy." Now these Banks will unload at the end of the level (when the BanksLoader GameObject is destroyed).

Try out the level, and it should still work. If not, double-check that you're definitely loading all the Banks you need.

Once you're happy that everything's working, open the GameOver Scene. Obviously this will also need a Banks loader, so repeat the same process.

This FMOD Studio Banks Loader component will obviously need to load the GameOverBank, and you can also set this to Unload on Object Destroy. However, does it require the MasterBank and MusicBank? Play the Scene, and you're going to see error messages.

Well, the answer is "it depends." If you play this Scene on its own, you'll get error messages. If, however, you play the Battleground Scene and stand in front of a SentryDrone for long enough, then the GameOver Scene will load without any issues, because the necessary Banks were never unloaded. At this point, you should have finished playtesting the Scene, so it's safe to leave these out.

Repeat the same process for the YouWin Scene (it will need to be exactly the same as the GameOver Scene – if you want to be clever, you could make the GameOver BanksLoader GameObject into a prefab, then re-use this in the YouWin Scene. This has the added advantage that you can then use the Prefab master if you later need to make changes to both Scenes).

That's going to help a little, but we can take things a little further – for example, we can unload the SentryDrone Bank as soon as all of the Drones are destroyed. Return to the Battleground scene, navigate to the MyScripts folder, and open the DestroyMeWithRocket script.

If you remember, this script is attached to the SentryDrones, and causes them to be destroyed when they collide with a Rocket.

Paste over the code from the DestroyMeWithRocket2.txt file, save the script, and return to Unity.

Next, we have to Tag the SentryDrones.

- Go to Edit > Project Settings > Tags and Layers.
- Create a new Tag of "Sentry."
- Select the SentryDrone prefab master (it should be in the MyPrefabs folder).
- Set the Tag to "Sentry."

Now, when the last SentryDrone is destroyed, the SentryBank is unloaded.

```
 1 using System.Collections;
 2 using System.Collections.Generic;
 3 using UnityEngine;
 4
 5 public class DestroyMeWithRocket : MonoBehaviour
 6 {
 7     public float sentryCount;
 8
 9     // Use this for initialization
10     void Start()
11     {
12         //print ("Sentry Count = " + sentryCount);
13     }
14
15     void OnTriggerEnter(Collider other) //Code is called when a collider or rigidbody touches the trigger
16     {
17
18         if (other.gameObject.name == "Rocket(Clone)") //Determines tag of objects that will destroy the object
19         {
20             sentryCount = GameObject.FindGameObjectsWithTag("Sentry").Length;
21             if ((sentryCount - 1) == 0)
22             {
23                 //print ("All Sentries destroyed");
24                 FMODUnity.RuntimeManager.UnloadBank("SentryBank");
25             }
26             Destroy(gameObject); //Destroy the collided with GameObject
27         }
28     }
29
30 }
```

Figure 28.2 DestroyMeWithRocket 2

Walkthrough: Unloading the SentryBank

We've used most of the elements in this code before, so you might be able to follow it on your own.

We start by declaring a float called "sentryCount."

In the OnTriggerEnter Method, if the GameObject that the script is attached to collides with an object tagged as "Rocket":

- We count the number of GameObjects tagged as "Sentry," and assign this to the sentryCount variable.
- If (sentryCount -1) is equal to zero, the SentryBank is unloaded. (The reason we're checking for sentryCount -1 == 0 [rather than sentryCount == 0] is because this section of code is run before the SentryDrone is destroyed.)

Loading and unloading Banks can make a game run more smoothly, but it's not always going to help – if you have a sudden number of Banks and Events that all need to load at once, this can lead to a logjam. This is something that will have to be determined by playtesting.

Data Compression

FMOD has the ability to use data-compressed files (e.g. mp3). However, there is no benefit to this. When your Banks are Built, FMOD applies data compression. Data-compressing an mp3 probably won't make it any smaller – and is very likely to make it sound significantly worse.

Go to the FMOD preferences (FMOD Studio > Preferences), select the Build Tab and click on where it says "Desktop" to access the settings. From here you can set the default data compression format. The options are:

- Vorbis: named after a character from a Terry Pratchett book, this is a data compression format similar to mp3.
- FADPCM: FMOD's own data-compression algorithm. Optimized for signal processing in FMOD.
- PCM: uncompressed.

253

As a rule, use FADPCM – especially if you're performing a lot of signal processing in FMOD.

If you need even smaller files, use Vorbis, as it has a higher data compression ratio.

The smaller a file, the quicker it can be loaded. However, it takes more DSP to perform signal processing on a data-compressed file – and if your files are small/short enough, then this can outweigh the advantage of compressing them in the first place. In which case, change to PCM.

It's also possible to set the data compression differently for each audio file from the Audio Bin. Open this up, and select one of the music files.

At the bottom of the Audio Bin is an outline of a blue dotted box. Click here, and you'll see you now have the ability to change its data compression settings.

If you set this to Vorbis, you have a Quality slider. Use the "Compress" button to A/B the original and data-compressed versions, and adjust the percentage so you can't hear any compression artifacts (bear in mind that you can probably take this a little further, as many of the compression side-effects will be masked in the mix).

FADPCM has a Sample Rate slider. Higher sample rates lead to larger files, but the limit to this is determined by the frequency content – the higher the frequency, the higher the sample rate needs to be. If this is set too low, you'll hear an unpleasant form of distortion called "aliasing." For low-frequency and atonal sounds (e.g. explosions) you might be able to get away with setting this quite low.

PCM also gives you a Sample Rate slider, but lowering this too far will give you a low pass filter effect (rather than aliasing). If your samples have no high-frequency content, then it's worth setting this as low as possible without affecting the program material.

Loading entire long audio files uses up more memory, as the file sizes are larger. We can avoid this by setting our audio files to stream from disk (streaming audio uses more CPU, so there's always going to be a trade-off between the two settings).

If you haven't customized the data compression, then there is a button next to the blue dotted box labeled "Stream," and unsurprisingly, it will make the audio file stream from disk.

If you have customized the data compression, you have the "Advanced Loading Mode." Here you can also set it to Stream, but you can also change the way that audio is loaded into RAM. Compressed will load the file as-is, but Decompressed will un-compress the audio back into PCM as it loads it to RAM. This won't give you any improvement in audio quality (once you apply lossy data compression, you can't get it back), but can reduce the load on digital signal processing. This isn't a setting that you'd often use – perhaps for a mobile phone game, where you have less available CPU?

Other things that you can do to get your audio running as efficiently as possible include:

- Re-use elements wherever possible. Perhaps elements of a punch sound could be modified and used elsewhere?
- Use mono audio files wherever possible. An uncompressed stereo audio file is twice the size of a mono file. Most diegetic samples don't need to be stereo – their positional information is given by their placement in the game world.

- Trim your audio files. Silence still takes up file size!
- Keep your files short. Does the ambience loop need to be 4 minutes long, or can you get away with something shorter?

Optimizing your audio is usually a series of small savings that gradually add up – it can be an art form in itself. While we don't have the time to go any further in depth here, it's something that you should keep in mind the entire way through sound design and implementation.

Now we're ready to share our game with the world, but to do this, we'll need to export it from Unity as a self-contained application. That's exactly what we're going to look at in Chapter 29.

29 Building the Game

256

Learning Outcomes

By the end of this chapter, you will be able to:

- Export your game from Unity as a self-contained application.
- Configure the application graphics and splash-screens.

Your game is now complete, but, so far, it will only be playable if you have Unity installed on your computer. In this chapter, we'll see how to "Build" the finished game, so that it is playable as a standalone application.

I've provided a number of graphics assets with the book resources, so import the entire GameLogos folder into the MyGraphics folder in your Unity project (though, of course, feel free to use your own!), then navigate the Project Tab so that you can view the folder contents. Almost all of these assets need to be set to be Sprites (and we can actually use sprites for everything we're going to do), so:

- Select all the graphics in the GameLogos folder.
- In the Inspector, set the Texture Type to Sprite (2D and UI).
- Click on "Apply."

This has set all of the icons as sprites at once.

Go to File > Project Settings > Player Settings, then click on the padlock at the top right to lock the Inspector view.

Player Settings are available for WebGL, Facebook and PC, Mac, and Linux Standalone. You can switch between these using their Inspector icons.

Build Platforms

WebGL

WebGL is a Javascript API that makes it possible to run games within a web browser, without needing any additional plugins. FMOD recently added support for WebGL – however, getting the two to work together is not yet very straightforward.

Facebook

Facebook's Gameroom allows you to publish directly to a Facebook page (this also makes use of WebGL). However, the Gameroom application is not yet available for Mac at the time of writing.

PC, Mac, and Linux Standalone

We'll focus on building as a computer application. It is possible to build for Mac on a Windows computer (and vice versa) – provided you checked the appropriate boxes when installing Unity (see Chapter 1).

Player Settings

The Player Settings allow us to set how the game behaves, as well as giving us the options to customize our logos and loading screens.

Company Name

This is set to "DefaultCompany," so you might as well change this. I'm going to use "Mega Excellent Superb Games."

Product Name

This takes the same name as your Unity project. This is likely labeled something like "AudioForGamesFMODandUnity," so take this opportunity to name it something more interesting. I'll call this "Bazooka Berserk."

Default Icon

This will be the icon for the application. Drag the GameIcon.png sprite into here from the Project Tab.

Default Cursor and Cursor Hotspot

Our game does not have a cursor, so this isn't relevant to us. However, if we did, this would allow us to set a custom cursor icon, and set its click-point (relative to the top left corner of the icon graphic).

Resolution and Presentation

Most of this is fairly self-explanatory. We've already changed the Run In Background setting to make it easier to use Live Update, but this also affects the final game build.

Capture Single Screen is only relevant to Windows builds, and prevents the game from dimming other screens in a multi-monitor setup.

Display Resolution Dialog determines if you can change the resolution settings when the game runs. The options are Disabled (you can't), Enabled (you can), and Hidden By Default (you can if you hold [alt]/[opt] when the game launches).

Splash Image – Icon

This allows you to override the game icon for the current platform. This can be useful if you have versions for different resolutions and icon sizes.

Splash Image – Application Config Dialog B

This allows you to customize the Resolution Dialog window, by adding a graphic at the top. This can be up to 432 by 163 pixels in size. Drag the ApplicationConfigDialogB.png sprite into here.

Splash Image – Logos

The option for "Show Unity Logo" is permanently checked – this is one of the few limitations of the free version of the software. All it means is that the Unity logo is displayed when the game launches.

Drag the BazookaBerserkLaunchLogo.png sprite into the Logos box, then click on the plus button underneath, and drag the MegaEx cellentSuperbGamesLaunchLogo.png into the new box that appears. You can then set the duration sliders to determine how long each of these will be shown for.

By default, the Unity logo will be displayed underneath the other logos, but you can add this to the sequence by changing the Draw Mode to "All Sequential." You can see what this will look like by clicking on "Preview."

Experiment with the Splash Style (how the Unity logo is displayed) and Animation options; see what you like…

If you drag the SplashBackground.png sprite into the Background Image box, you can add a (blurred) image behind the launch logos.

There's not a lot that will make much sense to you under the Other Settings category, and the XR Settings pertain to virtual reality, so we're pretty much done here.

File > Build Settings, and the Build Settings appear in a floating window (you can also see there's a shortcut button to access the Layer Settings from the Inspector).

You will already have three scenes included in the Build, but if you've added any more, you should add them here. The scene at the top of this list will load first when the game is run. As we've optimized our Bank loading, this is going to be very important to get right.

Next, you can select the platform that you're building the game for. Several of these options are only available if you have a license (for example, you can't just build a game for PS4). Leave this set to PC, Mac, and Linux Standalone, and choose Mac OS X or Windows as your Target Platform.

There's a check option for Development Build. This gives you a build of the game that can connect to Unity's "Profiler," which can monitor and record the real-time performance of the game. While this can be useful, the game file will be significantly larger, so leave this unchecked.

Click on the "Build And Run" button, and you will be asked where to save the build. If your Target Platform is Mac OS X, then you can save it wherever you like. A PC build results in two files, so it's worth creating a new containing folder at this stage.

Click on Save, and Unity will make your game. The first time you do this it will take quite a while, so go and have a coffee break... (subsequent builds will be much quicker).

Mac Builds

Everything you need is packaged in a single .app file. Double-click on this to launch the game.

Windows Builds

A Windows build consists of an .exe file and a Data folder (the .exe is the actual game application, but it references everything included in the Data folder). Double-click on the .exe file to launch the game.

What's Next?

That's it! You've finished your first game – so what's next?

Well, the obvious thing to do is to jump right back in and make another one. Try seeing how much further you can take the game and sound design. Why not go for a totally different style? How about an FPS game where you play a toy fighting monsters under your owner's bed? Think outside the box, and try and come up with something new and original. We've already seen that Unity comes with various vehicle and character assets, so make the most of these to give yourself new audio challenges.

Expand your experience with other Game Engines and Audio Middleware – there are free versions of Wwise, Fabric, and Unreal available.

Most importantly, start working with other people and game developers – don't wait until you think you're ready – start right now! A great opportunity to gain experience and make contacts is at Game Jams. This is where people come together over a weekend to make a game. These events are open to people with any level of experience, and there's usually a shortage of audio specialists there (many game developers won't know which end of a microphone to point with, so you'll be appreciated).

Then (as ever), it comes down to persistence, time and effort – oh, and don't forget to enjoy what you do!

261

Index

265